backyard farming

ann williams

prism press

Published in 1978 by

PRISM PRESS
Stable Court,
Chalmington,
Dorchester,
Dorset DT2 OHB

Copyright Ann Williams 1978
Illustrations copyright Colin Browning 1978

ISBN 0 904727 67 X Hardback
ISBN 0 904727 68 8 Paperback

Edited, designed, typeset and made up at Prism Press

Printed in Great Britain by

UNWIN BROTHERS LIMITED
Old Woking, Surrey.

BACKYARD FARMING

contents

acknowledgements

Nearly all the farmers I have ever met have contributed something to this book and much of the basic theory was drilled into me by the staff at Studley College in Warwickshire where I studied for the National Diploma in Dairying.

For specific advice I am grateful to many, among them are Michael Armstrong for legal advice, Mrs May of the British Goat Society and my brothers David and Peter.

To my parents and to Colin Spooner of Prism Press I am very grateful for their encouragement.

Ann Williams

1 introduction to backyarding

There is no satisfaction quite like that of growing your own food. Complete self sufficiency died with the Dark Ages and nobody would want to be on that treadmill again, but on the other hand complete dependence on outside sources for something as basic as food means a loss of control over the family's life and even health.

This book is intended as a guide for those who would like to know what they eat, and have it fresh, uncontaminated and economical. There is no mystery about keeping animals and growing food for them; it was once everybody's province and it is our misfortune in this age that we have travelled so far from natural things as to forget them. There is now a need for some technical background for those who want to get back to home food production and this I shall try to give from my own experience for the most part.

I hope that you will enjoy reading about backyarding and that the book will help you to plan your moves or will explain something you have noticed; but I would not suggest that it can take the place of experience because no book can ever do that. Don't be discouraged if the skills do not come to you at once. It takes time and patience to learn how to make good cheese and butter, for example. There is a 'feel' for such processes that will only come with experience. Some books fire the enthusiasm and make things sound easy, and the reality can be disappointing. Things can go wrong; I hope that the information in this book will help you to cope when they do! Worthwhile projects are seldom easy. Take it gently, one thing at a time.

The Benefits of Backyarding

These you have probably worked out already; they are several. Food is increasingly processed and artificial. Farming techniques lead progressively to more adulteration of the food at source. The best way to get good food is to produce it yourself. It is also a means of saving money; partly from tax, because to buy food you have to earn the money and pay tax on it first. There are no fortunes to be made from backyard food, but it is a good way of recycling all kinds of wastes and turning them into something useful. Pigs, poultry and rabbits can turn weeds and food scraps into meat and eggs. Surplus produce can be bartered for other things you need.

Once you start producing, the quality of your fresh food will put up your standards so that you find shop eggs or milk even less appetising and you will wonder how we have come to this in what passes for a civilised world.

Limitations

The main ones will be the time and space available — and of course, your own inclinations! Space can be slightly elastic with care. For example, how about keeping bees on the roof? Their flight path would be far above people's heads and they would find their own food, in town as well as country. Many of the suggestions in this book will need land, but not all. The art of the possible lies in starting now, rather than waiting vaguely for the future.

Time in which to do these things is a problem, because working with animals is a very regular job, particularly when they are kept for the most part indoors and they depend on you for all their needs. Even wide ranging sheep must be inspected once a day. The secret is to get the whole family or group involved, with responsibilities clearly allocated.

If possible, it is a nice idea to use backyarding as a means of keeping one member of the family at home. The money saved in a year could easily equal a salary and unless your job is a vocation, you would probably get more satisfaction out of running a few animal or garden projects. And for women with small children who need to be at home, backyarding is a useful and pleasant hobby. Animals can be kept when there is no one at home during the day, but some, such as goats, will need managing very carefully in this case.

Resources may be limited, but take heart; this is not a rich man's hobby. Backyarding is a state of mind and can start with bean shoots on the windowsill. Many houses have some space at the back for chickens or rabbits.

The Background

The growing of food to eat is the one thing that all the peoples of the world have in common, except perhaps those who gather it from the wild. All over the world, since the first cultivators got busy, land has been in demand. It has often been difficult to get hold of; it has been taken away from the poor and helpless by those with power. History is full of it.

Exploitation of land is not new, either. Those who farm now with no care for the state of the soil or the wild life are just the same as their remote ancestors who roamed the wild, exhausting a plot of ground and then moving on, thus starting the world's deserts. A third of the world's inhabited surface is now desert or nearly so; every continent is affected. Researchers say that there are 45 reasons for desert conditions and that 35 of them are caused by man. So mankind has a poor record, although it is only recently that we have begun to realise it.

It is a very good thing that the new generation of backyarders have a proper feeling for the land. Attitudes among those actually working the soil have naturally varied, and the traditional mixed farms of Europe were, in general, run on sound lines which put the soil first, above all other considerations. The last twenty years has seen this delicate balance upset by new techniques, but in the new wave of enthusiasm for growing things there is rejection of bad ideas and a wish to get back to the organic foundations of good farming.

The preservation of the cycle of natural materials, through growth, death and decay back to rebirth, is known as the organic cycle and farming which preserves this and does not try to short circuit the cycle is known as organic farming. If you read the literature you will find that exponents of the system have a great respect for ancient Chinese agriculture, which supported a high population for centuries by a thrifty habit of returning all wastes to the soil. Reading the accounts of this makes one wonder why we throw out our wastes and then worry about pollution. Organic waste of any kind will enrich the earth.

In the last century, all farming was organic because the short

3

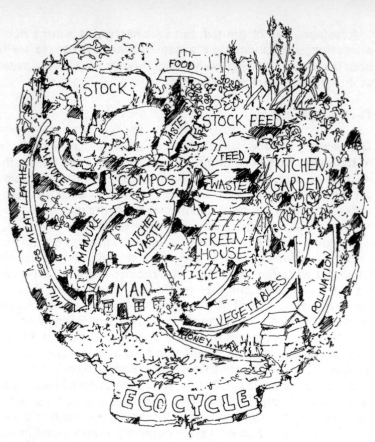

STOCK

FOOD

STOCK FEED

FEED

WASTE

COMPOST

WASTE

KITCHEN GARDEN

MANURE

KITCHEN WASTE

MANURE

GREEN HOUSE

MAN

EGGS MEAT LEATHER

MILK

VEGETABLES

POLINATION

HONEY

ECO CYCLE

cuts had not been adopted. Since all farming is to some extent
unnatural, the exact definition of organic practices is open to
argument, but the general principles are agreed by most people.
They are to try to avoid the use of poisonous substances, on
the land and for animal treatment, especially those poisons
which accumulate instead of breaking down into harmless
chemicals. This is important for plant and animal life but also
for the life of the soil, where micro organisms are an essential
part of the cycle. The idea is to try to build up a healthy fertile soil and provide natural plant and animal food, thus encouraging natural resistance to disease. This should make the
use of poisonous sprays unnecessary. Promotion of quick
growth by large applications of fertiliser tends to upset the
natural balance. On a backyard scale, it will not be needed. We
can afford to grow food for taste and quality rather than sheer
quantity. And if we do not use chemicals we will save money!
There is now a general feeling towards a kinder treatment of

4

the land and care of all natural resources, which comes under the heading of conservation; this may be an overworked word, but it is a good sign that we hear it so often. Careful use of our resources such as energy and water will come into the backyard plan.

It may be that globally, the warning has come in time. It is suggested that the spread of the Sahara southwards at its present rate of three miles a year could be halted by a belt of trees fifteen miles deep. It is therefore up to us to look after our own little patch; even on a very small scale, trees can be important aids to conservation. Lack of trees is leading to a world short-age of water. In part of Southern England once covered in trees, the water table (level in the soil below which all the pore space is filled with water) on the hills is now two hundred feet lower than it was; it has dropped since the trees were felled.

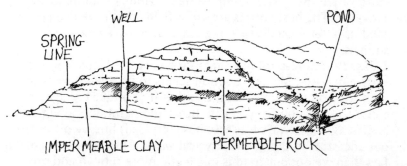

In arid zones, trees actually enhance rainfall; they also bring up minerals from the depths of the earth and improve soil structure by draining and aerating the soil with their roots. The water table fluctuates, and this can be seen when the pond as above fluctuates. In Britain it is usually highest in February and lowest in September.

We need to preserve or plant hedges as well as trees. They are little reservoirs of forest, shelter for wild life and they act as wind breaks for our stock and crops. This keeps the fields warmer and can sometimes help to keep the land from drying out too much. In areas without hedges, topsoil sometimes dries out and blows away.

Vegetarianism

Many backyarders are vegetarians and they point out that grow-ing milk and eggs is more efficient than producing meat. I hope they will find plenty of scope in this book. Diet may be a matter of principle, but for those who would like to eat meat if they could do so with a clear conscience, backyarding does solve some of the difficulties. Here are some of the arguments.

1. *Inhumanity.* Mass production in farming is repelling; hens in battery cages, veal in the dark. Cheap food policies encourage this sort of thing. Home production in small quantities ensures a supply of food from animals that have lead healthy, natural lives.

2. *Biological Inefficiency.* They say you can feed twenty people with the corn needed to produce meat for one, and in this unequal world, many feel guilty about eating meat for this reason. But not all land will grow corn. Grazing animals can convert grass, which we cannot eat, into meat, which we can. Also, organic principles demand some kind of livestock farming, the more varied the better; compost needs animal manure as an activator. The best corn is grown in fields that have been ploughed up after several years in grass, grazed by a mixture of animals.

3. *Health.* Cholesterol in animal fat has been blamed for degenerative diseases. Medical opinion seems to be mixed; work in California on the subject suggests that if you take enough exercise to keep you fit, you can digest any natural food without ill effects. This type of disease has increased, but we differ from our ancestors in doing less physical work. They ate more animal fats than we do; our food is in general more refined and processed. When you produce food at home it will not be processed, and also the hard work you put in will help to keep you fit enough to enjoy it!

4. *Economics.* Backyard meat production need not be wastefully expensive. Food costs are the biggest item in animal rearing, but as we saw before, many animals eat weeds and waste products. Garden and wayside plant material is always available in country districts. In the towns, there are household scraps from neighbours and food shop waste, the greengrocer's outside cabbage leaves, stale bread from the baker.

5. *Technological Inefficiency.* Yes, this is a grave charge on modern farming. They tell us that six times more energy goes into producing battery eggs than comes out. Home produced food is not likely to be so wasteful because most of the energy comes out of the eventual consumers, so you get a recycling effect! On a small scale there is no need for high energy inputs.

6

If you would like to rear rabbits for meat but are afraid of killing them, somebody will no doubt show you how until you get used to it. It is a problem for complete beginners, but the way of doing it quickly and painlessly can be learned and there need never be any suffering for your animals at all. When you have several animals, the breeding stock will be the ones you know and the offspring will be more or less anonymous when they are grown.

For committed vegetarians, a few animals will help the vegetable growing and there are many possibilities without growing meat. A goat or two for milk, some hens for eggs and a hive of bees for honey are some ideas you might like; in addition, the Angora rabbit can be kept for its fine wool, as can a few sheep if you have the room. Spinning and weaving are good backyard hobbies and you can grow your own clothes.

One word before we start, about recording. Records are a great help; perhaps a general diary would be a good start. It helps you to learn as you go along. To know exactly how long a goat has been in milk, how many eggs have been produced in a year, adds to your knowledge and experience. Time passes quickly when you are busy and it is easy to forget what has happened. With a record, you can learn from success and failure alike. This applies both to crops and animals.

2 planning

Never think that planning is a waste of time; it is very important
indeed. Planning is fun in itself, and it helps you to foresee
snags. It saves all kinds of trouble and inconvenience and animal
rodeos.

The first step is to consider your holding or garden object-
ively, its good and bad points. Maps will help; study maps of the
area including a geological map if you can get one. Make a map
of your existing layout, to scale if possible, and make sketches
of any proposals.

Think of the overall scope, which varies with every case. The
main point is the interrelationship of everything — crops,
animals and people. Such things as soil type, rainfall, elevation
above sea level, slopes and growing season are your limitations
or advantages to be worked with as they are; they can't be
changed. Some things can be altered; the fertility of the land
can be improved; buildings can be bettered, so can roads and
water supplies. Shelter belts can temper the wind and the whole
landscape can be altered by tree planting.

Soil Type

There are thousands of soil types and it is usual to find several
on the same holding, which is one of the reasons why some
fields will seem to do better than others. If you inherit 'bad'
land, try to work out why it is bad. Poor thin soils can be
improved by the addition of humus. They may carry few

animals at first, but as fertility improves they will support more, and erosion will be less likely.

Heavy clays can be cold and boggy, but they are also improved by adding humus, which is decayed vegetable matter, because it lightens them and allows them to drain better. Clays hold moisture, which is useful in drought conditions; light sandy soils are quicker to warm up in the spring, but dry out more easily. (See the soil chart).

Depth of soil is also important; if there are only a few inches overlying rock, the land is best left in permanent pasture which will build up feritility in those few inches. Soils overlying rock are the first to dry out in spells without rain.

Try to work out the soil type if you are inspecting a holding with a view to buying it; mole hills will show you what is there underneath and some unpromising fields can show quite good soil in a mole heap, which would indicate the possibility of improvement.

The soil type will have a great deal of affect on the planning

Acid Soils

SOIL CHART

Type	Advantages	Disadvantages	Crops
CLAY Diameter of particles less than 0.002mm 'heavy' soils	Hold more water Rich in potash Trees grow well Often small farms and fields with hedges	Wet in winter; poach if animals are left on land. Deficient in phosphates Should not be worked when wet, so operations can be curtailed by weather Slow to warm in Spring	Often left in permanent grass which is not grazed in winter. Best arable crop winter wheat; autumn seed bed preparations often easier than Spring. Mangolds and cabbage do well, sugar beet and potatoes may be hard to harvest. Best to include grass in rotation.
SAND	Can be worked at any time. Usually free draining, warms up early in Spring. Stock can be wintered outside because little risk of poaching.	Liable to drought in day periods — do not hold water. Low in plant nutrients Leaching occurs — plant foods washed away. Lime often needed in small amounts. May 'blow' when dry	Good crops depend on plenty of water and plant food being available. Often used for market gardening because such soils are 'early'. Where rainfall low, rye, carrots sugar beet and lucerne do well. On good soils barley, peas, potatoes also. Trees are usually conifers
LOAMS (intermediate between clay and sand)	Warm up fairly early Are reasonably drought resistant	Should not be worked when wet	The best all round soils. Any crop can be grown where there is plenty of depth. Mixed farming usual — arable crops in rotation with grass, dairy or beef on the grass

CALCAREOUS SOILS (derived from chalk and limestone rocks)	Usually free draining Will not usually need lime	Usually deficient in phosphates and potash. Very few hedges. Some still have no water laid on, no natural streams. Organic matter breaks down rapidly.	Wheat on deeper soils, barley, sugar beet, potatoes, kale. Only very poorest are left in permanent grass.
SILTS	Sometimes very fertile and should be easy to work.	Bad drainage — the soil tends to block up drains. Silt particles have no chemical reaction, so cannot combine with lime	Arable difficult. Best left in grass. Deep rooted plants will improve drainage e.g. lucerne, chicory, yarrow and other herbs.
PEATY SOILS	When reclaimed, soil breaks down to release nitrogen	Deficient in phosphates and potash. Old tree trunks surface and have to be removed, as soil level falls due to breakdown of organic matter. Badly drained, often arid	Potatoes and oats where crops are possible. Grassland often good, but must not be overgrazed or weeds will come back
BLACK FEN	Contain a lot of humus Rich in nitrogen	Blows away in dry Spring when there is not enough crop cover to hold it down. Cured by applying clay or deep ploughing. Poor in phosphates and potash. May lack minerals, e.g. copper	Usually all arable, wheat, sugar beet or vegetable crops. Might be better with some grass to hold it down

Where soils are free draining, and slopes are towards the sun, it may be possible to keep cattle outside all the year round, as of course is the practice on such soils where the climate allows. Heavy soils in wet areas, to take the other extreme, will 'poach' — churn up into mud — under the hooves of animals in winter. The soil as well as the climate decide what can be done; where land poaches easily, pigs, cattle and even poultry will need to be either very wide ranging or to be housed indoors for some of the time.

Acidity and alkalinity of soils is inherent, but can be adjusted. You may remember from gardening books that a pH of 7 is neutral; most crops prefer a soil at just slightly the acid side of 7, 6.8 or so. Soils over granite or sandstone tend to be acid,

ALKALINE SOILS

those over limestone are usually alkaline. The natural vegetation is often a guide; plants like sorrel, heather, mosses, bracken and foxglove grow in acid soils. Campion, chicory, toadflax and traveller's joy are plants found in lime soils. Ground limestone can be applied to correct lime deficiency, but I am told that a good organic soil can often look after its chemical balance and will not need lime. There are soil testing kits available from gardening shops, which should be used when the soil is moist and warm for a true reading. The state of the crops may tell you all you need to know, and sometimes the animals are a good guide. There have been cases of young cattle picking at mortar in old walls and this indicated that they were short of lime.

Rainfall

No year ever seems to be average, but figures will be available to give you an average for the region and these should be studied when looking at the overall plan. Some crops need more water than others; in areas of high rainfall, housing of animals and even equipment is important and hay making may be difficult.

It is possible where rainfall is high, to depend on it for a water supply, provided it is collected from the roofs in clean gutters, regularly cleared of leaves, and run into covered storage tanks. I knew a farm in a high rainfall area which happened to be on the top of a hill, with no water supply other than the rain. So rainwater was properly collected and the regular samples showed that it was clean enough for drinking. The quantity was enough for a herd of milking cows. Water storage is an important part of the plan; even where mains water is available, rain water will be a cheaper substitute for animal use.

In low rainfall areas, storage is even more important; it may influence the capacity of the holding, being used for irrigation in the drier months. There is another way in which you can plan to make up for low rainfall; get the soil into good condition. A good fertile soil retains moisture. The use of mulches is another aspect of the same thing; cover the ground with organic material and you cut down water loss by evaporation.

Elevation

This will also determine your scope. Going up a hill is like going away from the equator — it gets colder and less hospitable, and the vegetation goes through the same phases. But slopes to the sun can compensate for height. Many of the holdings taken by backyarders are on hillsides because these are the places not so suitable for commercial farming, or for building land. And hillsides can be good places to live. Drainage will be less of a problem, and air drainage is another thing to remember — cold air is heavier than warm air and it drains downhill, lying in the valleys in cold pockets, at night; during the day the valley will be warmer.

In a new area, it is always best to see what the neighbours in a similar situation do, until you have been there long enough to form your own opinions. I once planned a garden on a hillside where there had been no garden for years, and found that the hill sheep ate the vegetables. A good planner would have fenced it first.

Growing Season

This is the time from the last real frost in spring to the first autumn frost, and it limits what you can grow to things that are ready in the interval, or are frost hardy. Devices like a walled garden can alter the picture a little, but greenhouses are the answer in cold areas; seedlings can be started off under glass to gain time; plants such as vines can be kept under glass.

Land Use

This should be the prime consideration. If what you plan is not going to be good for the land, think again. Overstocking or keeping too many animals is bad; it can cause erosion, because bare soil with the vegetation eaten down to the roots is easily washed away by heavy rain. You can see this happening in areas of steep slopes and thin soils. On the other hand, in neglected fields without enough stock to graze them down, bracken and gorse come creeping back, spoiling the land for grass. With too few animals the grass grows coarse and tufty and is left by all but the pigs. Proper management and the right balance take time to achieve, but it pays to be flexible. A variety of stock will help. One example of good management is to let the cow graze the first fresh grass and let sheep follow, to tidy up. Then give the patch a rest before grazing it again.

Rotations are essentially a part of good land use; different crops have different needs and a variety will not deplete the

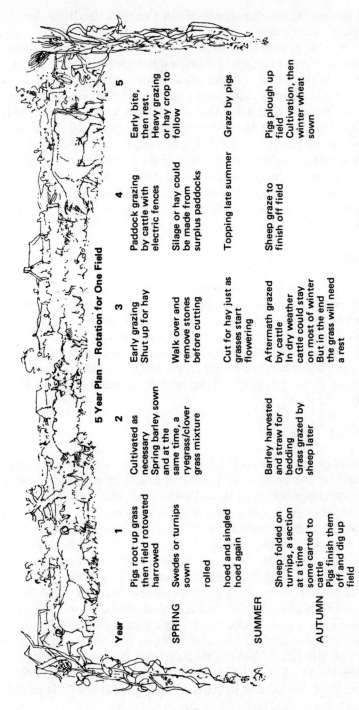

5 Year Plan – Rotation for One Field

Year	1	2	3	4	5
SPRING	Pigs root up grass then field rotovated harrowed Swedes or turnips sown rolled hoed and singled hoed again	Cultivated as necessary Spring barley sown and at the same time, a ryegrass/clover grass mixture	Early grazing Shut up for hay Walk over and remove stones before cutting	Paddock grazing by cattle with electric fences Silage or hay could be made from surplus paddocks	Early bite, then rest. Heavy grazing or hay crop to follow
SUMMER	Sheep folded on turnips, a section at a time some carted to cattle	Barley harvested and straw for bedding Grass grazed by sheep later	Cut for hay just as grasses start flowering Aftermath grazed by cattle In dry weather cattle could stay on most of winter But in the end the grass will need a rest	Topping late summer Sheep graze to finish off field	Graze by pigs Pigs plough up field Cultivation, then winter wheat sown
AUTUMN	Pigs finish them off and dig up field				

NB could leave down grass for longer if it suited the holding to do so. The mixture should be chosen accordingly.

15

soil too much. Gardeners will know that this also helps the crops to avoid pests.

It will be as well to keep as much variety as possible, a thing that commercial farmers are finding themselves unable to do. Grass, arable, wood and wet place all have their uses. It is a pity to sweep it all away, and clear and drain the lot in the interests of maximum production. Don't go all out for enormous yields; that is a path which has led farming the wrong way. True, sometimes drainage may be needed; some trees may need to be felled. But think before you change the nature of the area, and try to imagine the consequences of the change. Even scrubby woodland may be giving your fields protection from the wind in winter. Waterfowl and fish farming may be better than drainage for a pond.

Access

There may not be much choice here, either; the layout will be there when you take over. When changes are planned, access must be considered as regards to the movement of animals and supplies round the holding. Roads, gateways and fences need to be the best places. Go through summer and winter routines in your mind; collecting crops, feeding stock, removing manure. A work study exercise of daily chores can save a lot of time.

Fencing

By one of those perverse laws, fences are never as good as they should be and are often a permanent headache. Keeping them in good order is expensive and takes time, but is thoroughly worth while. There is a responsibility waiting for you as soon as you start keeping animals; they must be fenced in under your control so that they cannot stray and damage other people's property, and likewise your precious grass and crops must be fenced against any stray animals coming to eat it off.

Some of your boundary fences will be your responsibility and some the neighbours' but it is always best to see that your animals are secure in any case. Stone walls make good boundaries, but you can't choose them; if you live in stone wall country you will learn from the locals how to build them up again. Sheep tend to knock them down, and so unfortunately do tourists in some areas. In parts of Wales, walls are built of stones

16

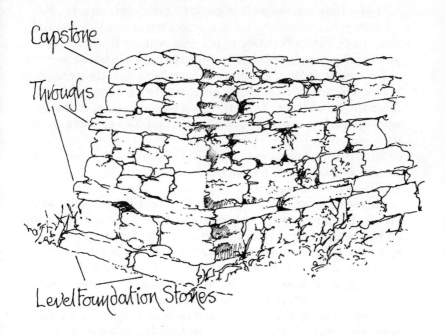

Capstone

Throughs

Level Foundation Stones

Section through a Drystone Wall

set in earth which makes them more secure for cattle, though sheep can run over them. The best ones have a hedge set on the top.

Stone walls need good foundations, and a course of 'throughs' every eighteen inches or so. These are flat stones projecting a little at each side and they give the wall stability.

Hedging is another old skill you may still be able to pick up in hedge country. If you have hedges, they should be kept trimmed regularly because if they are left too long uncut, they will grow straggly and will not keep in stock. Savage cutting down is not good; the best shape for a hedge is broad at the base and narrowing slightly towards the top. The bottom if it is nice and thick will harbour small animals and wild flowers and deter most stock except perhaps goats. They say that you can tell the age of a hedge from the number of different plant species in a given length. The more variety there is, the older the hedge. Some hedges in Britain are very old indeed; they are mentioned as boundaries in ancient parish records.

17

If you have no hedges it might be worth planting one along a fence. Hawthorn is the usual species, but it needs protection from stock for its first few years; they will eat it when it is young. This means two fences with the hedge in between.

Fencing is another art, and the secret lies in getting the posts in line and the wire taut. Oak posts will last the longest but larch will be cheaper. Treat the ends with creosote, if not the whole post. Some people boil them in creosote to make them last longer.

Shoots from coppices make light fencing if you weave them around vertical posts, the shoots about 2 inches apart, woven while they are still green.

Moveable fences are very useful for immediate repairs, windbreaks, temporary pens and so on. We have a selection of wooden hurdles, like light gates, which are useful for all kinds of jobs. The old fashioned wattle hurdle is a very good idea. Plant your own willows and in a few years there will be rods for hurdle making. Recommended varieties for this purpose include *Salix viminalis* mealy top, *Salix trianda* Blackmaul.

To make hurdles you split hazel or willow rods, about 8ft long, and weave them in and out of uprights or 'sails' which are set on a curve. At the ends the rods are bent back in again and

finish with the end inside. The split rods are knocked down with a mallet from time to time, rather as weaving is pushed down to keep it tight. This is a form of weaving with wood.

Old tyres can be used in many ways. We have made them into indestructible pig troughs and some people sole shoes with them. A strong man's art is making tyres into fencing. The tyre is cut across, the wire with a tough hacksaw and the rubber with a sharp knife, kept wet. It is then stretched out to make a continuous line which is threaded onto short poles. Two lines of tyres are said to make a pig proof fence.

Electric fencing is mobile and probably the most useful of all for internal fences. It enables you to make small plots of grass out of one field and to give the animals a change of grazing and each plot a rest. One strand is enough for cows, at about 3ft 6 inches from the ground. It is carried on metal posts with plastic or rubber insulators, and the posts have a pointed end which is easy to stick into the ground. A battery supplies the power, unless you have one which is designed to run from the mains. For pigs, two strands of wire are needed. There is now a special electrified sheep netting to keep in sheep; goats are a problem where fencing is concerned. An electric fence for goats needs to be somewhat elaborate. Sheep netting is a problem with horned animals.

Housing

Planning to house stock will depend firstly on what is already there. Some old buildings can be very well adapted for present needs. Permanent buildings are very expensive, but temporary ones will often be better. The simplest is a straw bale hut, protected by weld mesh, which is a well tried method of pig housing and can also be used to give shelter for lambing or calf accomodation. When it begins to look scruffy, it can be burned and replaced. This is a good way of controlling disease.

Another idea is to recycle unwanted materials, which are sometimes easy enough to find. They need not look like junk if the job is done in a workmanlike way. We once bought some holiday chalets for pig housing. The publishers of this book use reject garage doors.

Water

Many holdings which are bought cheaply turn out to have

water problems — either too much or too little. Lack of water is probably the worst. To find out whether there is a source of water on your land you could get hold of a water diviner — in Britain the British Society of Dowsers will tell you where to find one. Local water boards will give helpful advice and so will the Ministry of Agriculture.

In Wales once our family had a well below the level of the house. We solved the problem with expert advice by putting in an electric pump in the well itself and a pressure tank on the same level as the house.

Wells should of course be covered and fenced from animals, as should any water supply. A private water supply of unknown origin should be traced to its source so that if there is any chance of pollution, you will know about it.

Recent pollution is detected by testing for the presence of coliform bacteria. These are not natural water organisms and will soon die out in water. Their presence means fairly recent pollution by animal effluent. It would be wise to get a public analyst to test for purity in case of doubt, but one test does not mean a safe supply. It only means that at the actual time of testing, this was the state of the water. The Ministry of Agriculture will test water on registered holdings.

Rats

Any planning of buildings and storage must take account of rats. Those who do not like the idea of putting down poison should be particularly careful that it does not become necessary. Rats increase rapidly in numbers where there is food available, and you can soon have an infestation on your hands. They do a great deal of damage to fabric; they can gnaw through wood. They eat and spoil an enormous amount of food, and doing so they spread disease. So it is extremely important to exclude them. Preventive measures are as follows:
1. Build ricks and granaries on rat proof piles; stone piers 3ft high, with a saucer shaped rim of metal sheeting near the top.
2. Concrete foundations to buildings, lay brickwork in cement and use galvanised iron sheeting.
3. Meal bins should be concrete or metal. If wood is used, the angles should be covered with metal sheeting.
4. Proper construction, protection and repair of drain pipes, ventilators and basement windows, to make each building rat proof.
5. Disposal of waste food and rubbish so that it is not available for rats.
6. Protection of the natural enemies of rats — some of them rare. Owls, hawks, buzzards, rooks, crows, ravens, seagulls, stoats, weasels, foxes and cats. (Twenty freshly killed rats were found in the nest of a Barn Owl.) Unfortunately, some of these creatures are also enemies of poultry or lambs!
7. Burrows from which rats have been driven should be filled with cement mixed with broken glass.

Remedial measures, when you've got them, are hunting, trapping and poison. Plaster of Paris mixed with flour will kill them and is not a toxic substance to have around; that is, another predator eating the rat would not die. But the mixture must be kept away from other animals.

Planning for Produce

The first part of this is planning for variety. Within the limita-

NUTRITION CHART

Food	Function	Sources
CARBOHYDRATES Sugars — monosaccharides glucose fructose galactose — disaccharides sucrose maltose lactose Starches — digestible polysaccharides Cellulose — indigestible polysaccharides	Provide energy and may also be converted into body fat Energy is the ability to do work: some is lost in heat	Energy reserve of plants is main source of carbohydrates Lactose in milk. Fructose in honey. Typical Western diet is now 1/3rd sugar and 2/3rd starch — sugar content of carbohydrate intake used to be lower. Potatoes contain 0.4% sugar, 17.6% starch Wholemeal bread contains 0.9% sugar, 45.8% starch
FATS Saturated fatty acids, e.g. palmitic acid stearic acid butyric acid Polyunsaturated fatty acids Monounsaturated fatty acids all plus glycerol to make a Triglyceride Solid at low temperatures but become liquid when heated	More concentrated source of energy than carbohydrates. Contribute to texture and palatability of foods. Digested slowly, so have high satiety value (make you feel full longer)	Energy reserve of animals and some seeds Animal sources include meat — beef contains about 17% fat. Cheese — about 30% fat. Fish — especially fatty fish e.g. herring, mackerel. Butter — 81% fat. Vegetable sources — seeds e.g. sunflower, peanuts, soya beans

PROTEINS

Amino-acids — about 20 used. 8 thought to be essential for adults: Isoleucine Phenylalanine Leucine Threonine Lysine Tryptophan Methionine Valine Plus histidine — essential for infants Quality of protein depends on its ability to supply all these amino acids. Mixed diet best	Essential constituents of all cells, to regulate processes or provide cell material; needed for growth and repair. Excess of protein can be converted into glucose and used to provide energy. In a diet with not enough energy, protein will be diverted to provide energy from its proper purpose.	No single food supplies all needs. Mixtures of foods complement each other and result in better values. About 1/3rd protein in average diet comes from plant sources and 2/3rd from animal sources. The amount of protein in nuts and dried peas is the same as in meat, fish and cheese. This is lowered by cooking in water. Cereals are rich in protein; wholemeal bread is 9.6% protein. Potatoes contain 2.1% protein when raw — roast, 2.8%

MINERALS

15 are known to be essential: Iron, calcium, phosphorus, sulphur, potassium, sodium, chlorine, magnesium, fluorine, zinc, copper, iodine manganese, chromium, cobalt	Three main functions: 1) Constituents of skeleton — calcium, phosphorus, magnesium 2) soluble salts controlling composition of body 3) Necessary for enzymes and indirectly concerned with release of energy (Iron and phosphorus)	Calcium — milk, cheese, green vegetables Iron — meat, bread, potatoes, vegetables Magnesium — green vegetables; deficiency rare Sodium and chlorine — none for babies. 4g common salt per day needed by adult in temperate climate

TRACE ELEMENTS

Cobalt, Copper, Chromium, fluorine, iodine, manganese zinc	Knowledge of their functions is incomplete. Chromium has something to do with using glucose. Fluorine associated with structure of bones and teeth. Iodine, manganese and zinc are associated with enzymes	Cobalt can only be utilised in the form of Vitamin B_{12} Tea, water. Sea food. Content in vegetables depends on the soil, so use iodised salt in areas where soil is deficient in iodine Plants e.g. nuts and whole cereals

Food	Function	Sources
VITAMINS		
A (retinol)	Essential for vision in dim light and maintenance of skin	Vitamin A in fish liver oils, kidney, eggs, green vegetables and carrots
B Thiamine B_1	For steady release of energy from carbohydrates	Milk, pork, eggs, vegetables, fruit, whole grains (cooking may cause losses)
Riboflavin B_2	"	About 1/3rd riboflavin from milk; also cheese, eggs, meat, potatoes, tea, yeast extract
Nicotinic acid Pyridoxine B_6	"	Meat, bread, vegetables and milk Particularly in meat, eggs and fish
B_{12}	metabolism of protein and formation of haemoglobin deficiency leads to anaemia and degeneration of nerve cells	Only in animal products as a rule. Exceptions are comfrey, yeast extract
Folic acid	deficiency leads to a characteristic form of Anaemia	
pantothenic acid Biotin	releases energy from fat and carbohydrates metabolism of fat	Animal products, cereals, legumes Offal and egg yolk
C ascorbic acid	maintenance of healthy connective tissue	Most from fruit and vegetables, milk, rose hips, blackcurrents, potatoes
D	maintains level of calcium and phosphorus in blood	Obtained from action of sunlight on skin, herrings, cheese, eggs
E Tocopherol	can be stored in body so deficiency unlikely	Most foods. Richest are vegetables oils, cereals and eggs
K	necessary for clotting of blood. Deficiency unlikely	Vegetables. Can be synthesised by intestinal bacteria

24

tions, what variety of food can we produce? I have included a nutrition chart just to refresh your memory on the theory of our food needs, but nobody can accurately lay down quantities needed. So planning for quantity will be individual to each family; as the work increases you may well need more food! It is not easy to estimate needs six months ahead. And yields are variable according to luck, management and the weather. In the various sections I will give an estimate of the yield which can be expected from the different enterprises and you will know how often you can face for example, rabbit for dinner. Planning for milk, you will realise that you can't drink the entire output of a goat, so there will be some surplus to make into milk products.

This leads on to the handling and storage of produce. Freezer capacity is a limiting factor in some cases, but there are many other ways of storing food. Old holdings have cellars and cool dairies and are much better equipped for storage than new houses; you will really need a room in which to make butter and cheese and for storage of such things. Cheese should be stored in the right conditions, and if you put the butter next to the paraffin there will be trouble. A really good store/workroom would be helpful.

Budgeting

In this case it means allotting the winter food supplies for all the animals, particularly where the winters are long. In theory you can always buy in some more food, but in practice this might be unprofitable. Whether you grow all the fodder or buy in some of it, it pays to know just how they will all be fed, as prices tend to rise over the winter. A simple way is to multiply the animal's daily needs by the number of estimated days on winter feed — which is 180 days in northern Britain. A large cow for example will need 20 lbs of hay a day for maintenance, so we will need well over a ton per cow if we feed nothing else. A ton will be enough for a Jersey. More succulent feeds such as kale or roots are a great help for milking animals in winter when there is little grass about. For goats, these could be grown in the garden.

Buying in hay will be essential for a garden project unless you can make it from the grass verges; it may be a good thing for a larger holding. In this way you will be able to keep more stock and their manure will build up fertility, so that in a year or two the land will produce more and perhaps the extra hay can be grown at home.

3 machinery and equipment

Alternative Technology

Faced with a choice of technology levels, most backyarders will feel the attractions of 'alternative' technology. Most of us share the general concern over the use of nuclear power; we are often the kind of people who worry about what will happen when the world's supply of oil runs out. We like the idea of living off our own resources and not being connected to anybody's grid, to be charged their prices and subject to their strikes. Some of us live in remote places where there is no electricity available.

There is now a considerable literature of alternative technology and a Centre devoted to its study at Machynlleth, in Wales. There are available designs for windmills and waterwheels, so that if you have an engineering bent it should be possible to fix something up for yourself. These things can be very expensive, but some of the plans specifically set out to use second hand materials.

Energy is the capacity to do work, and is all ultimately derived from the sun. Of the total energy used, 62% is in the form of heat and 38% in the form of power. Over the world as a whole only 1% is used by agriculture, in spite of its reckless use in factory farming! It is interesting to note that per head of population, the West uses a minimum of 116 kilowatt hours per day; the Third World uses 9 to 30 kilowatt hours.

So it would seem that in terms of energy it will be more efficient to do things for ourselves where we can, as primitive societies still do; to use hand tools where possible. Backyarders, producing food on a small scale, can more easily do this than commercial farmers. But as individuals our power output is small. Compare the power of man, 50 to 100 watts, with the horse at 500 to 750 and wind and water mills at 2,500 to 20,000! A horse is ten times as powerful as a man, but it needs a skilled man to make it work.

Solar Power. Sunnier countries are obviously at an advantage here; North America usually gets more sun than Britain. Roughly speaking, the south and east half of Britain gets about 1600 hours of sunshine a year, the north and west about 1200 hours. The USA varies widely, but the average seems to be about 2400 hours for all but the extremes. It seems though, that a solar collector is worthwhile for about 1000 hours a year, so most of us may qualify. There is also the possibility of using solar radiation on cloudy days.

One way is to use solar heating intermittently, as it occurs, with an alternative way of heating the house or the water when the sun goes in. This is what most of the British installations are equipped to do.

The simplest solar panels are just like radiators; old water radiators, painted matt black, can be used. They are exposed to as much sun as possible, which warms the water in them. As it gets warm, it rises into the storage tank. The inlet pipe is high in the tank and below it is the outlet, for cold water to flow out to take the place of the warm water. Fix the collector so that it faces south, perpendicular to the noon sun. Insulate the back with polystyrene, cover the front with glass and leave about an inch gap between the glass and the radiator for the heat to build up. The circulation has to be shut off at night.

The principle of solar heat can also be used to cool things, and enthusiasts have put a solar unit in a conventional gas fridge.

Wind may be the best energy source in some places in the world. It is one we used to use, but have almost forgotten about, witness the derelict windmills scattered over the countryside. Denmark is perhaps an exception because the Danes, being short of some other sources of power, have kept on using windmills and therefore have kept them up to date.

Wind speed increases with height, so choose a high spot or a tall windmill. Dabbling in wind power, quiet and clean, is a fascinating hobby and there are several successful home made units about, but there are dangers attached to it. If sails fall to pieces at speed in high winds, they can be lethal.

Winter winds blow harder than summer winds, which is a

good thing since winter power needs are greater. A most useful
kind produce electricity which is stored in batteries for use
when it is needed. Otherwise, one has a vision of everything
running down when the wind drops!

Water Power. With a stream on your property this may be an
idea to play with; although any tampering with water runs you
into contact with legislation because in Britain you do not own
the water as it passes through your land!

Even quite a small stream can be useful, and the first thing
to do is to estimate the width, depth and speed of the stream
and therefore the rate of flow, before you go to talk water
power with someone who knows about it.

The old mill by the stream probably had a vertical wheel,
with the water passing either under or over it. Just occasionally
you may find an old one and be able to restore it, but for new
installations these have long been outdated. The newer types,
less picturesque perhaps, are turbines or impulse wheels. The
big problem, as ever, is to get a fall of water for the necessary
force; either a dam or a mill race is needed.

Oil. We are so used to oil and most of our machines are geared
to it, that it takes a determined effort to ignore it. Most back-

yarders will accept the need to use some oil, albeit with care. Paraffin is cheaper as household fuel than electricity and many a little holding is warmed, lit and fed by paraffin, as was the case last century. Even when there is electricity available, it is not always where you want it; storm lanterns are useful when tending stock on a winter night and as an alternative for when the current is switched off. Paraffin pressure lamps give a very bright light. The Tilley lamp is a mobile pressure lantern, a very useful tool. The use of oil for tractors we will consider later. *Methane.* There is a great deal to study in this subject. Methane is a gas produced by the action of bacteria without air, working on sewage sludge, and it is used in the same way as any domestic gaseous fuel. The residue has had nitrogen added to it during the process, and so it is a very good organic fertiliser. The whole thing sounds like a dream of efficiency and recycling.

However, there is a snag. Heat is needed to keep the thing going, and unless there is manure available in large quantities, there is almost as much energy input as output. The best temperature is 95°F which is why it works well in hot countries.

Animal Power

The Horse. There are points for and against the use of horses for power. On a small scale, the expense is not justified. William Youatt, writing in 1843, says about the working horse: 'The annual expense of a horse depends on:

1. The interest of purchase money
2. Decrease of value (he means depreciation)
3. Hazard of loss
4. Value of food (the biggest item today. An average sized horse if there is such a creature, will need about 3 acres of grass all the year round, but horses and cattle go well together)
5. Harness, shoeing and farriery (harness is very dear indeed. The blacksmith, if you can find one, will need to visit every six weeks or so even if the horse is not shod — unless you can trim the hooves yourself)
6. Rent of stabling — or shall we say what else could you keep in the stable?
7. Expense of attendance — or the work and time involved. Would it be quicker to do the jobs by hand?'

All these points are still true and it really depends on whether, if you have the land, you like horses and know them. If you have worked with horses you have a power tool ready to your hand; if not, there is so much to learn that I think lack of knowledge, combined with expense, makes the horse a doubtful proposition. I have worked with horses as well as with tractors; a wise old horse taught me how to turn hay. But well trained, wise old horses in farm work are difficult to find these days. They cost a great deal to buy, and to keep.

Ponies used for riding can be trained to pull a cart for odd jobs and this could be very useful. For light work like this, just occasionally and not too regular, they can live on grass, with hay in the winter and an occasional handful of corn. But horses on grass are in soft condition and will not be fit for real work. To do serious work a horse needs real food — up to 12lbs of oats a day. To grow the food or even just the grass for a horse will mean the loss of a more productive animal. And then of course, once you have a horse that is 'corned up' and fit for work you have to keep him working. Otherwise he may become unmanageable, or may develop some ill such as laminitis, from having too rich a diet. On a smallholding it would be hard to keep the horse ready for work when it was wanted.

To go a step further backwards in time, we could perhaps consider the ox. The French Charolais cattle are spreading round the world because they produce beef so well; but their enormous muscles have developed because traditionally they were used as draught animals. They are docile and a bit dim, but it should be possible to break them in for the job that their ancestors used to do. It would be marvellous to have a Charolais cow to produce beef and milk and to pull a cart. Milk production would not be high, but it might be enough for one household.

The drawback now I think would be to find someone who knew how to handle working oxen. Horse experts are rare, but the last team of oxen in my part of the world walked into the sunset about 1910 and I have so far failed to find anyone who knows how to work them. The harness and way of working are different from horses. There are still bullock carts in some places in the world, such as Africa, and it is to these places we must look for knowledge. I would be very pleased to hear from anyone who uses cattle for draught.

Tractors

These again need some specialised knowledge. It is no use at all to buy an old tractor and then to pay for repairs, and hours of time can be wasted by breakdowns and failure to start. Even on a farm scale, I find myself avoiding a tractor where possible. In this case, though, the necessary knowledge is more accessible. There are often evening classes on tractor maintenance and metal work and things of this kind, so that all you need is a basic interest. If you are a good mechanic, tractors will come easy to you; cheap ones are rare, but an old one, properly run, will be a great help. But without the inclination, don't get one — backyarders can often manage without. There are several ways to avoid using tractors; pigs to plough, the humble barrow, contractors for field work and helpful neighbours. Or of course a garden implement with power and two wheels, which will cost less to run.

Of course, ploughing is not the only job; in some places it may not be done at all. It is the heaviest work and it would not be sensible to be fully geared up for the occasional ploughing. Other jobs, perhaps more important to be able to do for yourself, are cutting grass or corn, and carting crops back to the buildings. Carting is nice work for a pony if you have one. Some people use a Land Rover or an old truck. American Indians and Yorkshire dalesfolk had in common a form of wooden grass sledge with which to drag home loose hay.

Tractors can be fuelled by petrol, tvo or diesel oil. Petrol tractors are expensive and are now rare. Older models tend to run on tvo — tractor vapourising oil. They have to be started on petrol because the tvo will only vapourise when the engine is warm. So there are two tanks and you switch over from petrol to paraffin when she gets warmed up. But tvo is now going off the market.

Tractors now are very versatile. They have a power take off shaft to turn machinery such as a muck spreader or a baler — or a water pump. This provides power for the machine as well, whereas tractors without a pto would be merely haulers, pulling things over the ground; and in the early days this machinery was land driven, that is, it was turned by the action of its wheels. Tractors now also have a pulley wheel which can be harnessed by a belt to drive all kinds of machines, from sawing wood to grinding corn. An old tractor can be very useful as a stationary engine.

The tractor hydraulic system is another very useful feature; oil pressure raises a pair of arms at the back of the tractor, which makes it possible to lift machinery clear of the ground for road travel and for turning at the end of a row. Small loads can be

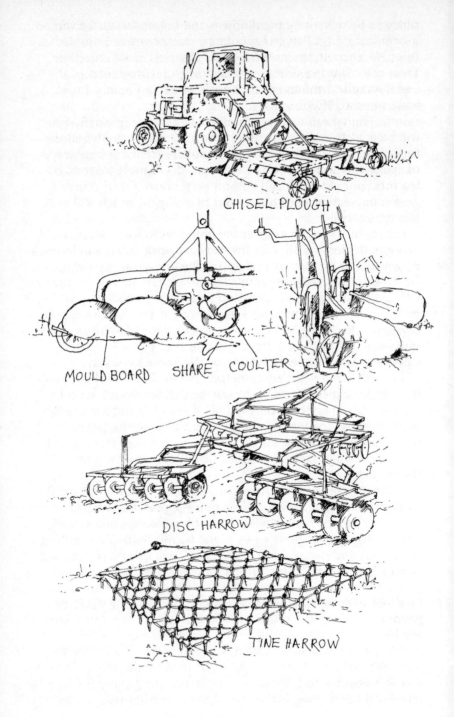

CHISEL PLOUGH

MOULD BOARD   SHARE   COULTER

DISC HARROW

TINE HARROW

carried on the hydraulics and the system can be used for working fork or shovel lifts to move large amounts of manure and so on. Most modern implements are designed to fit onto the three point linkage system of a tractor and this consists of the two hydraulic arms and the top link, which is a rigid support and acts as a kind of brace.

A difficulty about agricultural machinery in Britain now is the proliferation of safety regulations, which get tighter all the time. They are designed to save employees from the negligence of employers and you can sometimes get away with less than the maximum requirements if you are an owner-driver, but then it puts any group activity out of the question, because it will be illegal for anyone else to drive your tractor unless it complies with the regulations. No doubt they are a Good Thing, but some who have learned to live with machinery and have a proper respect for it don't feel the need for such elaborate precautions. Safety frames now have to be fitted to tractors so that if you turn it over on a slope, it won't crush you. This frame at the time of writing costs about £100 and since it will be a standard fitting in future, the price of a second hand machine will go up by this amount. Other regulations call for covers for moving parts such as drive shafts and gears.

Tractors are tending to get bigger and heavier, which makes their sheer size unsuitable for backyarding. It also brings the problem of impaction of the soil. The weight of such a heavy machine travelling over the ground, especially in wet weather, can flatten and destroy the soil structure, with its pockets of air necessary for bacterial life. This is an aspect of modern farming we will do well to avoid. The actual power (expressed in brake horse power) of a tractor is not so much more than a car; for example, the little Ferguson tractor equals a Mini car.

Garden Tractors

There is now a very good selection of garden tractors on the market, geared to the needs of backyarders. To buy a new farm tractor for backyarding would be out of the question, but the best buy in new (and thus more reliable) machinery would undoubtedly be one of these; the prices vary widely, but so do the possibilities. At one end of the scale is the simple cultivator, with which you walk along. We have one of these for the vegetable garden because although we have a farm tractor, there is never enough ground clear at one time to make it worth using a big machine. The cultivator speeds up digging and takes the backache out of the job. The new models are no doubt easier to handle than our old one; I find the noise and the effort makes them harder work than digging, but we do cover the ground more quickly. It will be wise to try one out before buying — the coloured adverts of happy laughing backyarders make it all look so easy.

At the top end of the price scale, the machines are miniature farm tractors and can perform all the jobs that a tractor can do; they have power take off shafts and three point linkage. They would be ideal for the work of a holding up to about 10 acres or so — if you could afford one!

Implements

The plough partially inverts the top layer of soil and buries the vegetation, which rots down to become humus. The parts of the plough are a curved mouldboard. Digger ploughs have short and steep mouldboards, the semi digger is less so and the general purpose is longer. Then there is a share, a coulter and a landside skim.

The share is the tip of the mouldboard and when it gets worn it can be replaced. This makes the horizontal cut and the coulter makes the vertical cut. Disc coulters are used where there is a lot of plant growth, knife coulters on hard ground.

Ploughing is a skilled job and it takes time to learn; watch an experienced ploughman and learn from him before you try it yourself. Much of the skill lies in 'setting out' before you start. Ploughing matches are the place to go, even though the fancy competition ploughing may not be quite what you had in mind!

After the land has been ploughed it stands in ridges, which need to be knocked down into a more level surface and a finer tilth before seed is sown. Machines for doing this are many; they can also be used for smoothing things out after pigs have been used for digging up the land. Cultivators are of several types, with pointed tines which travel through the soil. These are replaceable like the ploughshares; they are fixed to a frame, either trailed behind or on the hydraulics.

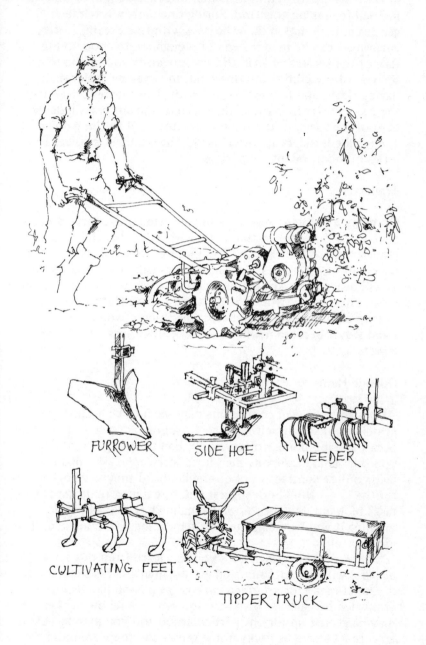

FURROWER   SIDE HOE   WEEDER

CULTIVATING FEET

TIPPER TRUCK

Harrows are lighter implements for the same job, or they can be used for raking grassland. Spring tine harrows are much gentler in their action than the ones with fixed teeth. Rotary cultivators can be used instead of ploughing, to chop up the surface of the soil rather than to turn it over. If you intend to do the rest of the cultivations by hand, this may be better than having ridges and furrows to deal with. The rotor is driven from the power takeoff shaft, so the speed of the cutters is independent of the speed of the tractor. For the final levelling down of the seedbed, rollers are often used. The Cambridge roll is a series of independent metal rings.

Trailers

On many a holding there are still converted horse carts; all you have to do is to replace the shafts with a drawbar. Ingenious recyclers make trailers out of all kinds of materials such as the back half of van bodies. Garden tractors will pull a small trailer, as will heavy cars. At haytime and harvest you will need a trailer most, but a van or truck will do almost as well. If we had to give up all our machinery except one thing, I would keep the Land Rover because it will do light field work as well as road travel.

Outside Help

Although the use of contractors may seem to be against the principles of sturdy self-reliance, it can be far cheaper to get the occasional job done professionally than to struggle hard to do it yourself. There are many men in country areas who make a living out of contract work on the land and they keep machinery for the job. I would not suggest you hire a man to do simple tasks by hand, but for things like baling hay, ploughing or ditching it will be much cheaper than hiring or buying a machine yourself.

There is another angle too; you can learn from this system; because with the machine you get the man and his expertise.

Sometimes a local farmer can easily do a job like baling for you when he bales his own hay; sometimes a fellow backyarder may be geared up for one particular job and you may be able to arrange a barter. As backyarding grows and there are more people with an interest in it, specialisation is developing a little. If you find you are good at shearing sheep, you could do this for your neighbour in return for his help with something else.

Hand Tools

Practice is all with hand tools; you buy the best you can afford
so as not to blame them for a bad job, in time the blisters go and
the muscles settle down and you find you are enjoying the job
and taking a pride in it. With a little skill, hand tools are effective. The price of labour has almost made them obsolete on
big farms, but it is surprising how much work can be done with
muscle power.

William Cobbett's Cottage Economy includes a precise des-
cription of how to keep a cow on a quarter of an acre, which you
dig yourself and plant with cabbages and swedes. With practice,
digging is fairly easy except in very dry weather. In an hour or
two a day a large patch can be covered; if there are several in
the family willing to help, you can dig a garden big enough to
grow all your own vegetables without using any machine.

Spades. Garden spades have a slightly concave blade, while
ditching spades are more hollow. They all need to be of good
steel so that they have a sharp cutting edge and a high polish;
this really does make the work easier. So you clean them after
use and put them away, to keep them in this efficient condition.
Forks. Forks for digging usually have four straight, flattish
prongs attached to a cross bar. Forks for lifting manure have
longer, more curved prongs; I have been using one all morning,
and would be much more tired had I used a garden fork, so it
pays to have the right one for the job. Root handling forks have
knobs on the ends of the prongs so as not to damage the roots.
Pitchforks have a long handle and two curved prongs, used for
hay and for corn sheaves, to pitch them onto a cart and off to
the stack.
Shovels. These are used for handling loose material — grain,
manure, sand. In some areas the handles are longer than in
others.
Hoes. Row crops are still hoed and singled by hand in some
cases even now. A good long hoe with a rectangular blade is useful if you grow turnips or mangolds. Keep the blade sharp for
cutting weeds.
Scythes. These are hard on the back, but mowers used to cut an
acre of hay in a day with a scythe! They are still used for awk-
ward corners or orchards, and if you get the knack of using one

you could cut your own hay or corn. The sickle is a smaller version, for one hand, very useful for cutting green food for rabbits or hens.

Wheelbarrows. The wheelbarrow replaces the trailer for small quantities. Get a solid but light weight barrow with a pneumatic tyre. Old wooden barrows look picturesque but are too heavy. Even better, a multi-purpose large barrow (but light weight) built on motor cycle wheels and axle, with a large bin on top — good for shifting hay and bulk crops etc.

Crowbars. These are very handy; they can be used for making holes in hard ground when fencing, or as a lever to shift heavy weights. A heavy hammer will be essential for driving in stakes. Lighter hammers and an assortment of wood working tools will no doubt be part of your equipment, depending on your skills. A damp proof workshop to keep them in will make the tools last much longer. Anyone doing demolition work will need a pickaxe! A tool collection is acquired gradually and as the need arises; the main thing is to keep them in good condition and to know where they are. Hours can be wasted looking for stray tools and it really does pay to have a place for everything.

Borrowing tools is a good way of finding out what you really need, but follow strict rules about it. We have found that people will be pleased to lend simple tools and equipment if you always return it promptly and in better condition than you found it.

Sometimes farm sales are a good source of hand tools and equipment; the prices vary but always know how much the thing costs to buy new, so that you have an idea when to stop bidding. Sales may also be good places to look for halters, ropes, troughs, racks and such things.

Grain Milling

There are now table models for enthusiasts to mill their own
flour. If you grow an acre or two of grain, you may want something bigger — say a mill powered by a tractor drive belt. Old
mills can sometimes be found at farm sales. Rolling mills, which
crush the grain between rollers — 'corn crushers' — are only
really useful for animal feeds unless you are hooked on porridge.
They will roll oats for horses and crack corn for poultry.
Grinding Mills. Ground or rolled grain is a more effective animal
feed and can be better utilised, but beware of grinding too finely.
Very fine barley meal is not good for pigs and fine meal is particularly indigestible to ruminants. Grinding mills are an arrangement of metal plates through which the corn passes; they can be
adjusted to the degree of fineness you want.

Hammer Mills. These are for large scale operations and can do a
lot of work in a short time. One advantage is that you can feed
whole oats or barley with the stalks on, unthreshed, into the mill
and get a meal that is part roughage. This would also save threshing operations.

Grain Storage

Grain should be stored in a dry place, in metal bins if possible.
Large farmers harvest their grain in damp weather sometimes,
and they dry the corn. If you want to keep yours without it

going mouldy it should be harvested at about 16% moisture or lower. We found it worthwhile to buy a moisture meter and harvest when the grain was dry. The main problem apart from damp is rats and mice — hence the bins.

Milking Machines

Those who like to keep things simple may prefer hand milking the house cow or goat, but there is something to be said for mechanised milking. Cleaning the equipment afterwards will take some time and may be rather onerous for the milking of just one animal, which is of course a drawback. On the other hand, it is much easier to keep milk clean when extracting it by machine, and clean milk keeps longer and is the foundation of good dairy produce. Hand injury may make milking difficult and a machine will be a great help in such a case.

Farmers are now for the most part milking in parlours or through pipelines in cowsheds, which means that old fashioned bucket plants can be bought at farm sales. These are usually fixtures but can be dismantled and put up again. The exception is the Gasgoigne Miracle Milker, a useful little gadget which is a self contained unit on a trolley and only needs an electricity supply. The principles of all the makes of milking machine are the same, and there is not much to choose between them in efficiency. I like Alfa-Laval and Gasgoigne the best.

The milk is extracted by the application of vacuum; about 15 inches of mercury or 7 lb per square inch, (half atmospheric pressure) is used to milk cows and a little less, say 11 inches, for goats. Constant vacuum would cause the blood circulation to stop, so there is a pulsation system which relieves the vacuum in the shells - the part between the rubber liners and the metal teatcups - and this makes the liners collapse on the sides of the teat and massage them.

The machine consists of a motor, which can be electric or a petrol or diesel engine, a vacuum pump which the motor drives; a length of metal pipeline making a circuit from the pump and back to it, passing over the place where the cow is milked; a trap at some point on the line, to stop any liquid going back to the pump; a vacuum regulator and a gauge to show the vacuum level. The recommendation for the engine is one horse power for four units, so a very small engine will be enough for one unit.

These 'units' are the actual machines, consisting of a metal
bucket with an airtight lid, a set of clusters ie teat cups, and a
rubber pipe to take the milk from the teat cups to the bucket.
Another pipe connects the whole thing to the vacuum line and
the bucket in effect becomes a vacuum reservoir. The pulsator
which sits on top of the bucket lid, causes intermittent vacuum.
The normal ratio of vacuum to normal pressure is 3:1 and the
pulsators tick at the rate of 45 per minute.

There are a few simple maintenance jobs to attend to, such
as topping up with oil, but the main chore is washing up and
milking machines can be run for years with no special mechanical skill, as I know from experience! The fact that most cows
are now trained to machine milking could be a good reason for
installing one for your house cow.

Special machines with just two teat cups are made for goats
but may be difficult to find. Cow machines can be modified by
blocking off two of the outlets at the clawpiece - the metal
part in which the cups converge - and this is often done by goat
owners.

I would advise anyone who buys a secondhand milking machine to replace the 'rubbers' with new ones. They perish when
not used and are very difficult to sterilise when old. Get new
teat cup liners and new milk tubes from the local farmers'
hardware shop; keep an old one as a pattern as there is a bewild-
ering variety of design.

4 poultry

For starting a backyard enterprise with rather limited capital
and experience, you can't beat poultry. They are small enough
to handle easily and fresh eggs are a real luxury; as small a unit
as half a dozen hens can keep a family supplied with eggs for
much of the year.

The subject of poultry keeping has been very well covered in
the Backyard Poultry Book, but we can consider the main points
of poultry management and also see how they fit into the overall plan. The main systems of keeping hens are set out in the
table. The strawyard or the deep litter system will be the most
suitable for small areas; those with a field or two can let the hens
range freely, as we do. We find that they can pick up part of
their own food, particularly the insects and grubs which might
otherwise prove to be pests! This is a very satisfactory method
of pest control.

If your backyard is in a densely populated area, it may be
better to keep the hens indoors, in which case deep litter is the
system. There is no need to consider keeping hens in batteries
at all, because such small birds can be given freedom to scratch
about, flap their wings and run about. I am quite convinced that
batteries are cruel; this summer we gave their freedom to a few
bought-in battery hens, and it took them days to learn normal
hen behaviour. They had hardly any feathers when they arrived.
Hens can manage with as little as 4 sq ft per bird, which works
out at a house 24 sq ft for six birds. There are still backyard hen
houses on the market, which you could either buy or copy.

Some people have managed to convert an existing building,

42

such as an old wash house, or put a loft over the garage. Requirements are fairly simple; a warm dry weather proof building, a few nestboxes, suitably dark and private; perches for them to roost on at night, a water drinker and perhaps a food hopper. Litter can be straw, dried bracken, hay, shavings etc. A free source of litter is the ideal, and in country areas grass can be cut from the wayside and dried, or bracken can be gathered from upland areas. Eggs are excellent bartering units, so once the hens are in production some arrangement about swapping eggs for straw or shavings will be possible. The principle of deep litter is to keep the stuff dry and friable, not wet and sour. If it is dry the hens scratch about in it and enjoy it, as their remote ancestors used to scratch about on the forest floor. Rake over the litter occasionally and break up any big lumps.

Hens like to perch at night because they used to fly up into the trees for safety; and some bantams revert to the old habit at times! A rail in front of the nest boxes for them to alight onto when they fly up is a good idea, but they perch on the highest rail at night, so make the roosting perches higher than this. Most of the droppings will be under the perches, so a moveable tray, cleaned twice a week, will help to keep the litter clean.

SOME POULTRY SYSTEMS

System	Housing	Feeding
Deep Litter	One house for everything: hens kept inside on dry litter which hens keep turning over. You add more litter from time to time. Suitable for loft, garage etc. No smell— will work in towns Nest boxes 16"x16"x 16". Perches 2"x2"	Scatter grain among litter. Put meal or mash in hopper. 4oz feed per bird per day. 3 oz grain 1 oz protein Hang up bunches of greenfood, to keep it off floor
Strawyard	Open yard, fenced with netting and littered with straw, plus wooden house for roosting and laying. House must be weatherproof. Perches and nest boxes as before	Green shoots can be provided by giving access to mown grass Waste vegetables and scraps can form part of ration.
Folding	Moveable wooden ark or henhouse on wheels with moveable run. It is trundled on to fresh grass every day. House for nesting and perching	Allow for fresh greenfood, amount to vary according to time of year, and also a few insects
Free Range	Fixed hen house with nest boxes, perches etc. and with access to large field or alternate runs. Gives a deep orange colour to egg yolks because of carotene content of green food.	Green food and insects will be available, also grit in most soils.

Other Requirements	Problems
1) Clean fresh water ad.lib. will need about ½ pint each per day 2) Grit for gizzard to help in grinding food— say ¼oz per bird per month 3) Calcium to help strong shells e.g. oyster shell Collect eggs every day	Litter must not go wet and sour or it will smell and birds will not be happy. Litter disposal if no garden (it will be welcomed by anyone with a garden)
As before; better to keep food hopper in hen house. Shut hens in at night. Provide dust bath — box filled with fine earth or sand.	1) Price of straw 2) It may blow away and make a mess elsewhere 3) Dogs, cats, foxes 4) Birds of the air steal food 5) Neighbours may complain of noise 6) Hens may fly out — but wings can be clipped
Must be level ground and predator-proof. Should not return to same patch too often. Makes a good treatment for grassland	The bother of moving every day
Plenty of clean water and the chance to get back into the house when they like.	Hens may lay in hedges etc. when you can't find the eggs. Foxes may get the hens if they are not shut up at night

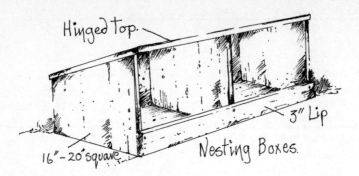

Hinged Top.

3" Lip

16" - 20" square

Nesting Boxes.

They will need clean water, so the fountain type of drinker is a good investment, as is a proper food hopper. Greenfood will be a nice change for indoor hens and it can be suspended just within reach, but clear of the floor.

Point of lay pullets will make a good start. These are young females about five months old, ready to start laying at any time. present prices are about £2 each so the outlay is not too high, and it is a quick way into production. Later, when you get used to keeping hens, you could try the delights and hazards of rearing your own chicks. Pullets hatched in spring will be ready to lay by autumn and will lay for about a year, with a little extra lighting in the dark days. By the next autumn they will stop laying and go into a moult, which means they lose feathers and look rather dejected. Commercial producers cull at this stage, ie they get rid of the hens and start with a fresh batch. (Moulting can in fact happen at any time after a shock or fright; anything that goes wrong can start it off)

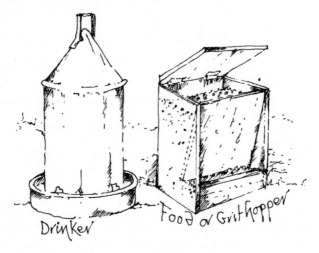

Drinker

Food or Grit hopper

46

Since a really high output of eggs is only obtained with young hens, even small scale poultry keepers often dispose of about half their hens in early autumn and replace them with pullets. This ensures a steady flow of eggs.

What breed is best? The table of breeds presents a summary of some of the possibilities. The hybrids are highly geared for egg production and they will demand a high standard of management. Personally I think that the older breeds are more suitable for backyarders. They are usually hardier and more likely to produce happily, at a lower rate perhaps, but on cheaper food. If you like chicken meat, one of the dual purpose breeds such as the Rhode Island Red will be useful. The cull hens will make a meal, and also if you hatch your own chicks, the cockerels can be fattened for the table. Pure breeds should last longer than hybrids, which are discarded by intensive producers after only a year. And also, pure breeds will breed true; that is, the off-spring will resemble the parents. Hybrids are not sold for breeding, they will not necessarily breed hens as efficient as themselves; you are supposed to go back to the supplier when you want some more pullets. This is also true of pigs and rabbits.

The other device used by backyard poultry keepers is sex linkage. Unless you are an expert, it is difficult to tell the sex of chicks until they grow, but with certain crosses the colour is linked to the sex, so you know what they are as soon as they hatch. This works when you use a red cockerel such as a Rhode Island Red with white hens such as white Sussex; all the girls will be ginger. It doesn't work the other way round, though. The advantage is that if you only want to rear pullets you can sell or barter the young cock chicks to someone who will fatten them.

A cockerel won't be necessary just for egg production; the hens will lay very well without one, but the eggs will not be fertile. Some people claim that fertile eggs are much better for you and taste better than infertile ones, but I have as yet seen no scientific evidence about this. The cockerel may be unpopular with the neighbours if he crows too loudly and too early in the morning, although some people will welcome such a country sound. But he will eat a lot and an extra hen in his place will give you eggs. It may be a good idea to start off without a cockerel, and to progress to one later.

Feed is the big problem with hens, because of the cost. As the hen's crop(the pouch in the neck where the food goes first, you can feel by handling it whether it is full or not) is not very big,

SOME POULTRY BREEDS

Breed	Advantages
Hybrids e.g. Kimbrown) Brown Arbor Acres S.L.) Eggs Hi Sex White) White Shaver S444 T) Eggs	Very good layers Economical — eat less food than older breeds Easily obtainable Won't go broody — a good thing in layers
Rhode Island Red 5–6½lb weight Dual purpose	Good for both eggs and meat Brown eggs — good to sell or swap Fairly placid Good mothers
Light Sussex 5½–6½lbs weight Dual purpose	As above. Also can get sex linkage with R.I.R. cockerel so sex of chicks is according to colour
Maran 6–7lb Dual purpose	Very nice large speckled brown eggs. Plump birds, good for table, attractive plumage
Leghorn) light Ancona) breeds 3½–5½lb	Economical — eat less, take up less space
Bantams miniaturised versions of all the old breeds — about half size	Less food needed Less room Very hardy Long lived Good broodies Some claim special flavour for eggs

Disadvantages

Intolerant of less than top class
management.
No breeding of replacements, will
probably not breed true
No broody hens for mothering chicks
May not be used to natural conditions
Will only lay 1—2 years

Eat a lot
Cockerels can be vicious

White eggs — not so attractive
Prone to excessive fat if kept indoors

Eat a lot!

Not much meat
May be flighty

Fewer eggs
Smaller eggs
May be broody too often
Rather flighty and nervous
Cockerels make a row

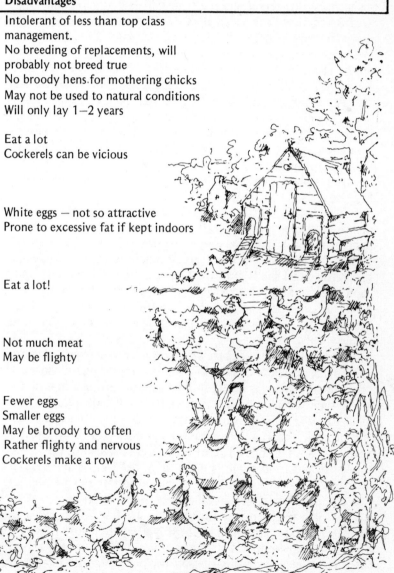

the food needs to be fairly concentrated when the birds are in production. The hen can accomodate about 4 oz of food at a time. Carefully balanced layer's meal costs a lot but it could be used for part of the ration. If you can find a source of protein such as insects and grubs, or fish or bone meal, grain is good as a main ration; scatter it in the litter and they will enjoy sorting it out. Wheat is the best grain for hens, and a cheap source of wheat for cottagers in grain growing districts is to glean like Ruth in the fields of Boaz. By this method we have fed our hens.

In the right area, permission to glean should be easy to get; of course if you have your own corn crops, the hens can be run on the stubble after the corn is carried, and can do their own gleaning. Few farmers run stock on the stubbles now, they are cultivated almost immediately after harvesting to minimise plant disease, so you have to move quickly. Get in behind the combine; there are no stooks now, but the big machines sometimes cough and when they do, a nice quantity is lost on the field. Nobody is going to get down from one of those machines to scoop up a little corn, so that's where you come in, with a fire shovel and a little bag.

Most combines pour the grain up a spout and into a trailer, but the old type put it into bags. Anybody who still uses a bagging combine will be able to let you have some cheap 'tail corn' for the hens. On this type of machine there is a special spout for weed seeds and small poor grain seeds which would spoil the sample; this stuff is ideal for poultry feeding, but the modern combines spread it back on the field, which no doubt is one of the reasons why corn fields get weedier and need more sprays.

The hens will enjoy corn wherever it comes from, for one meal a day, and the old backyarder's standby was mash for the other meal. This was household scraps, potato peelings and anything nutritious boiled up and dried off with bran or layers meal to make a nice crumbly mash; it should of course be cooled and not fed hot. It smells so good you feel you could eat it yourself. With an open fire or solid fuel stove, kept going all the time, the hen food can be boiled up at no extra expense. Waste bits of weed and vegetable from the garden or the greengrocer will complete the diet of your happy hens.

The main thing to avoid with backyard hens or ducks is the dirt run. A small grass run in front of the hen house may seem a good idea at first, but the grass gets eaten down and the soil turns to mud or beaten earth and the hens peer disconsolately out through the wire. It has been calculated that to keep a run green each hen needs 25 sq yds, and of course this is a waste of space that could be used for something else.

What is the answer? Some people use two runs alternatively, letting one rest while the other is in use. Keeping them off the land altogether where space is tight may be the answer, which is why I mentioned deep litter. But a limited free range in the garden is often possible; in the early spring, with tender shoots about, the last thing you want in the garden is a marauding hen, but in late summer and winter they could wander about, say for a limited time each day, and probably do more good than harm.

Breeding Your Own

If you decide to rear your own replacement pullets, a cockerel will of course be needed. Get a good one; he will be happy with about 12 hens. Then you will have to wait until one of the hens goes 'broody' if you want to do it the natural way. She will sit on the nest all day, talking quietly in a particular way. Wait until March or April so that the chicks hatch in spring weather.

About twelve or thirteen eggs can be hatched by a broody hen; an old superstition said always set an odd number of eggs. Get a nice little coop ready for the operation before you begin; a box for sitting should be about 16 inches square and about 30 inches high at the front, sloping to about 24 inches high at the back. The detachable front should let in little light, but plenty of ventilation.

Cut a large turf to cover the floor, about 4 inches thick, and put it in the coop upside down. This will supply the necessary moisture during the sitting. Scoop a shallow hollow in the turf and line it with hay, and try the hen on her new nest for a day or two to see if she sticks tight. Sometimes the fact of moving out of the hen house and into the coop is enough to put her off.

If she stays on the nest, introduce the eggs by stealth at night; give her whole corn, fresh water and grit, and plenty of quiet. (Do not allow the children to inspect.) Let the hen off the nest every day, for food. Some are so keen on their job that they have to be lifted gently off; be sure that they go back.

After 7 days some people candle the eggs to see if they are fertile; if held against a bright light, the infertile eggs will show clear, but the embryo chick will cast a shadow. This was done because after a weeks' incubation, the infertile eggs are still fit to be used, whereas if you wait until the hatching date they will have gone bad.

In hot weather the last week may be rather dry for the eggs, and they can be sprinkled with water. Watch out for chicks by the 20th day; if some come out before the others it may be as well to take them away and keep them warm until they are all hatched. Leave the hen alone with the chicks for a while to get used to them.

Hen and chicks can live in the broody coop, with the front removed and access to a run of fresh ground. Give them a chick drinking fountain and make sure that they drink. 'Chick crumbs' is the food for them; leave them with the hen about 7 weeks and by then they will be independent.

Another way to get to this stage is to buy day old chicks, which can be given to a hen to rear. In this case, have her sitting on dummy eggs for about a week, so that she will think she has done the whole job herself. Put one chick under the hen at first and leave it for a while, then introduce the rest gradually, in the evening if possible. Remember to shut them up at night in case a rat steals the chicks.

Artificial brooders can be used for rearing instead of a hen if you are starting from scratch or have no broody. Chicks need a temperature of 98 to 100°F but this can be arranged with an infra-red lamp. Use a thermometer and adjust the height of the lamp to get the right heat; a cardboard barrier is enough at first to keep them from wandering too far away. If they huddle together, they are cold. Chicks can manage without food for 48 hours after hatching, but they should be fed if possible before this. They can eat from shallow baking trays; from about the age of three weeks, household scraps can be introduced to the diet.

Eggs

Say 200 a year from each hen — quite a lot of eggs to deal with. They will keep in a cool larder for about four weeks, except for the cracked ones which will not keep so long. The important thing is to know in what order to use them, and not to get old and fresh eggs mixed up.

Store them pointed end downwards on trays, and keep them away from strong smelling substances. If you keep eggs in the fridge they will not last so long when taken out.

Surplus eggs can be stored in waterglass for at least a year. 'Putting eggs down' is an old country custom. Waterglass is soluble sodium silicate, made up by mixing equal parts of the powder with water or the paste according to directions. It can usually be bought at chemists' shops. This stuff seals the pores of the shell and the eggs are good cooking quality when they emerge. Store them in the solution, in an earthernware crock if possible; top up the solution as it evaporates.

Deep freezing of eggs is possible; the yolk tends to thicken, but the whites can be frozen just as they are. Freeze separately or together. Beat them up lightly and add a little salt or sugar to prevent thickening — and remember which! Or they can be frozen in an ice cube tray, one egg to each tray, and then the frozen cubes can be bagged. This will make it easier to follow recipes.

Eggs can be sold to householders without any restrictions, but NOT to shops in Britain.

Some Egg Faults

Soft shells	normal at start of laying season; check feed otherwise
Thin shells	not enough calcium, give oyster shell grit
Cracked eggs	thin shells, or no bedding in boxes, hens disturbed
Whites runny	old eggs, stored too warm, or very fresh eggs
Yolk pale	not enough green food
Blood spots	small haemorrhages

Geese

Geese are a very good way of starting with livestock if you have a little land, but not much capital. They will graze down grassland and eat grain on stubble; they enjoy wild and garden waste greenfood. It is particularly convenient if you have a stream or pond.

We made a mistake, I think, when we started with Chinese geese. They are small and the flesh is yellow and they were also vicious and attacked me frequently when I went to feed them. Beware of geese if you have young children around, but if you need watchdogs, they are recommended.

Better than the Chinese, which are small white birds, would be the other two old fashioned breeds, the Embden and the

Toulouse Goose

Embden

Toulouse — the latter is grey and rather attractive. Most of the geese you see on village greens are crosses of these two breeds, plus the Roman.

A breeding set is three geese and a gander, and thus equipped you could rear a few goslings. You could rear your own breeding geese from goslings if you can be sure of the sex, but the gander should obviously not be related to the geese.

Goslings are usually advertised at about the end of April and at a fortnight old they will be able to manage without heat and will be happy in a little grass run, with a shelter to sleep in. They make the oddest liquid warbling sound when going off to sleep. Grass and chick crumbs or bread and milk will do to rear them, plus plenty of water in a deep container so that they can immerse their heads, which is important. The grass for geese should be short, so put them on a mown patch.

Once they are out of the gosling stage, they can manage on grass alone until it is time to fatten them with barley meal in the autumn. Laying birds will need a ration of about 5 oz oats daily, or swill; we fed our birds on boiled potatoes, stale bread and scraps. Your breeding set should produce at least 100 eggs a year between them, and the surplus can be sold or bartered quite profitably.

Make a straw nest in the shelter they use at night to encourage the geese to lay there, but keep a look out in case they make a nest outside. Eggs are laid from mid February onwards; getting them hatched can be quite a problem because not all geese will go broody. The Chinese we had never seemed to go broody at all.

The eggs will keep without being sat upon for about ten days, which gives you time to think of something. Mark them with the date and turn them every day — this is important as the contents should not settle. A broody hen will be able to manage about five goose eggs. The loan of incubator space is also worth looking for. Surplus eggs can be eaten; they are large and have a strong flavour.

Geese are long-lived and will be a good long term investment. They will love a stream or any stretch of water, which they will of course fertilise with their droppings. The fox is their biggest enemy and so geese should always be fastened up at night.

Ducks

There are many advantages to keeping ducks. If you have a fishpond, they will help your backyard fish farming; the pond will be sealed by their flapping feet, it will be fertilised and the weed will be kept down.

Ducks' eggs are bigger than hens' and the Khaki Campbell even lays more eggs than the average hen. Ducks are rather hardier than hens and less prone to disease; they do well on poor land. The eggs are laid at night or early morning so if they are shut up in a house every night, the eggs will be there and you won't have to hunt for them.

Ducks do not need nest boxes or perches, so the hut can be very simple. They can be kept without water if they are able to immerse their heads in the drinker, but they enjoy water so much it seems a pity to deprive them of it. Given a pond or stream, so much of their food they can find for themselves.

Ducks do not make very good mothers. It is usual to let a hen rear the ducklings, although she becomes neurotic when they take to the water!

As we saw before, the Khaki Campbell lays the most eggs if you want duck eggs. For rearing ducklings for the table,

Khaki Campbell & Aylesbury

Aylesbury or Pekin ducks will provide white flesh. They are usually killed at 6 to 7½ lbs live weight, which they reach at 8 or 9 weeks, if fed well. This is the stage when they are exactly right for the table, and if kept longer they will cost more and not be so good. Kill before the pen feathers or proper quills emerge and they are much easier to pluck.

Before we leave the subject of poultry, there is one type of bird, much neglected for the table, that I would like to mention.

Pigeons

Pigeon keeping may seem slightly antisocial, because the birds may be suspected of eating your neighbour's garden peas. But they are very good for self sufficiency — the production of natural food in a small space. You can keep them in a loft and give them their freedom to fly about without needing any land; and they will be just as happy in a town as in open country.

Table pigeons called squabs, are eaten young, before they

fly; they are a popular dish in some places, including the USA. In Britain they were eaten at the beginning of this century, and the idea is coming back because you can now buy the table breeds. They would be a very good source of meat in an emergency. With no land, the urban backyarder has made a cult of fancy pigeons and fancy rabbits. Why not breed the same creatures for meat? Practical Europeans have been doing this for generations.

The best breeds for the table used to be the Carneau and the Mondain. The Carneau was rather hardier. It came from Flanders where it was bred for the table. A pair of these pigeons would rear about six pairs of squabs during a season. The Mondain was a bigger bird and very often white; it is still available. It has feathery feet, which has been considered a drawback because the quills can sometimes puncture the eggs when the female is sitting on the nest.

Housing. Pigeons for the table are often kept in a loft and not allowed out, but if you give them a flight each day they may be able to find some food for themselves. They should be kept in for a while when they first arrive and allowed to see out, to get their bearings. Once they have settled down it will be safe to let them out, before a feed so that they come back when you scatter corn.

A lean-to aviary 10 by 4 ft and 6 ft high would be big enough for six pairs of birds, about 4 ft of the length would be covered over with boards to make a nesting area — pigeons nest in lockers. These are shelves with a central aperture and an alighting board, projecting in front for them to land on. The shelves are about 18 inches apart, a foot wide and 4 ft long. There are two earthernware nest pans in each locker, half filled with sawdust, and sawdust is sprinkled on the locker floor — pine sawdust if possible since it discourages parasites. There are two nest pans because pigeons often lay a second pair of eggs while still feeding the first youngsters.

The droppings from the pigeon loft are valuable; they can be used in the process of tanning hides, and pigeon dung is supposed to be the best compost activator.

Flat perches about one inch wide are suitable for pigeons. The outside part of the loft could have a solid roof and wire netting sides.

Feeding. February to September is the time when the eggs are laid. Like hens, pigeons can be induced to lay in winter with artificial light. Put only mated pairs in the breeding pen; the best way to start will be to buy mated pairs.

Pigeons lay two eggs at a time and they are hatched after 19 days. The young ones are fed by the parents and they are ready for the table at 4 to 5 weeks old, when they weigh about 1 lb. They are killed and dressed like poultry. A loft with six pairs in it might rear about sixty squabs during a season.

Only the adult birds will have to be fed directly. A suggested diet is a mixture of equal parts of wheat, peas, millet and maize, the last just in cold weather. Crumbled waste bread and biscuit waste can also be fed. Other grains can be given to pigeons if they are available, but maize makes them too fat if they are not allowed out.

Grit and calcium are needed as for any poultry. There are special fountain drinkers made for pigeons, but they also need to bathe. A heavy wide shallow pan will be best, given to them for a time and then removed, every other day in summer and twice a week in winter.

Most town pigeon fanciers these days are keeping their birds for racing or showing; but rearing them for the table is a practical alternative and I should be interested to hear from anyone who does this with success.

The firm of Abbot Bros. who sell these types of pigeons tell me that some people keep them for laying the eggs, but use another breed for hatching them out. They will supply details on request.

A Few Words about Turkey Rearing

Turkeys are big business now, and many people feel that they can only be reared in very artificial conditions and on imported food. It is true that the climate of Britain may not suit them to perfection, but other types of poultry have also come from different climates and have adapted successfully.

For backyarding I think we must forget about factory farming; if we like the taste of turkey, it should be at its best if we go back to turkey rearing as it was in Norfolk in the 1920s. Fifty years ago, the birds were allowed to live outdoors in order to keep them healthy.

Disease has always been a problem with turkeys, but as usual, the fundamental cause is often management; too much

58

as well as too little. The Norfolk rearers felt that too much
care could be lavished on young turkeys. "Many of the most
successful rearers are gamekeepers, in whose methods there is
no fuss or sentiment", they say severely.

They had rules for the job which must be as valid now, and
as useful to know, as they were fifty years ago.
1. It is best if fresh ground can be used each season, ground
which has been free from any kind of poultry for several
years. They are kept in coops outdoors with a hen at first.
The problem here is that if they are too far from the house,
predators may carry them off. Also, they are rather silly
about taking cover in heavy rain and can die if they stay out
and get chilled; although one would think the hen would see

to this. The Norfolk men felt that the risk of shutting them up in a stuffy coop was greater, and they gave them free range from the start. One way was to put the coop in a walk cut through a standing crop of grass or lucerne.

2. Feeding was very carefully done. Free range meant they could forage for themselves, increasingly as they got older. At first they were given good stuff; chopped hard boiled eggs (or scrambled with milk) were mixed with fine biscuit meal (scalded) and chopped onion tops. Fresh whole or skimmed milk was always given. The chicks were fed little and often; more or less ad lib. Later their diet was extended to include cracked maize, boiled and mixed with middlings, ground oats or barley meal. Sometimes maggots were specially provided. Fresh meat offal was boiled for them and they were always allowed a lot of green food. Nettles were highly valued; they were boiled and the water (nettle tea!) used for mixing with dry meal to make a crumbly mash.

This sort of attention is unheard of now commercially; we do not cook food for our animals or make nettle tea. For backyarders, it would certainly be worth a try.

3. The strictest rule is no overcrowding. Eight turkey chicks to one hen in one coop are about right, and this would be enough unless you wanted to produce a surplus for sale. The mother hen can be left with them for as long as she is willing to look after them. Later, when they have outgrown the coop, the turkeys are encouraged to roost in a tree in the open, out of the reach of foxes. Disease was found to follow when they were housed.

5 rabbits

Rabbits are the best meat producers of all; and in addition, the Angora produces wool for you to spin. Rabbits are fast growing producers of meat of very high protein — only chicken is higher. The absence of fat means that the meat is digestible by children and invalids. Under good management they are happy in hutches so even with literally nothing but a back yard, good food can be produced at an economical rate.

Most commercial rabbit producers are now using expensive, usually imported grain in the form of pelleted feed, especially made up for rabbit feeding. But economical meat is produced in backyards, by long tradition, from greenfoods, hay and household scraps. Rabbits relish most harmless weeds. On a natural diet, they may take longer to grow than commercial ones, but the meat could be tastier as well as much cheaper.

The table of breeds gives you an idea of the choice for meat production. Fur is an interesting sideline, and the rex varieties have been bred with fur production in mind, although they are also useful for meat. The rex rabbit was a mutation first spotted in France; they have short, dense fur like velvet with no guard hairs (the long ones which project furthest). A later variation produced the satin, the hairs of which have a sheen.

Then there is the Angora, useful as I said for meat and wool; these are docile animals which are exhibited at shows in full magnificent coat, but it seems rather a job to keep them clean. The ones in production have a shorter coat because they are sheared or plucked every six weeks or so and the wool is spun with a little sheeps wool to produce a lovely soft yarn. The yield

SOME BREEDS OF RABBIT

Breed	Advantages	Disadvantages
Hybrids 12lb wt.	Easily available. Bred for rapid production of meat. Surplus rabbits should be easy to sell	Many not be hardy — reared indoors. May not last so long as older breeds. Will not breed true — you can't keep your own breeding stock.
New Zealand White 12lb wt	White fur and flesh which is favoured commercially. Large rabbit, has large families. Easily obtainable.	Rather large bones. An albino, so may not be very hardy.
New Zealand Red 11lb wt	Smaller bones than white. Good mothers. Easily obtainable.	Colour is a disadvantage if you want to sell surplus to packers or dye the skins.
Dutch 5½lb wt	Small, early maturing breed. Eats less than larger rabbits. Hardy outdoor type. Should be easily obtainable.	May be too small for family meals. May be too attractive!
Californian 11lb wt	Mostly white (with black tips) so good commercially. Fast grower, good temperament. Dense fur.	Not so hardy for outdoors in Britain.
Rex varieties of many eg Chinchillarex Havanarex 9lb wt	Short dense fur means you can go for skins as well as meat. Very fine quality meat. Most of them hardy and long-lived.	Darker meat than from commercial rabbits. Smaller carcase.
Angora variable size	Good quality meat. Produces wool — a clip 5 or 6 times a year. Can be kept for wool alone.	Scarce at the moment. Must be kept very clean. Some strains not good meat producers.
English 8lb wt	Good meat and not too much bone for an old breed.	Slower growing than commercials.

can be anything up to about a pound of wool from each animal in a year, but the American Angora seems to be a higher yielder than the British one.

The system of rabbit keeping you choose will depend on how much room there is available — see the table of systems for the alternatives. Perhaps the best compromise would be to allow them access to grass in the summer months, say in a Morant hutch on the lawn, and then to bring them into warm hutches in the winter. Trouble may be experienced with winter breeding if they are too cold. This will also allow the grass to have a rest which is important.

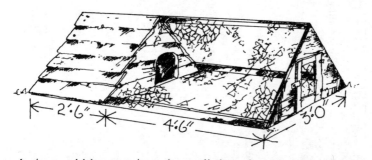

Indoor rabbits can do quite well, but there is more danger of respiratory infections, caused by breathing ammonia fumes from the urine, unless you are most particular about ventilation. A disused garage sometimes makes a good rabbitry, if it is light and airy. It is useful to have them under cover for your own comfort when attending to them, in places where the winters are severe. They will also be safer from the risk of the dreaded myxomatosis infection, which could be carried to your rabbits by the fleas on stray dogs or cats. Flies can be kept away and shade provided in summer, and the place can be kept rat and mice proof.

On the other hand, lack of an indoor setup need not mean your rabbits are deprived. Outdoor hutches are snug and warm and the fresh air makes for a healthy rabbit community. The best backyard hutch is probably a sturdy wood affair, standing outside on legs, weather proofed with roofing felt and with overhanging eaves to keep out the rain. It will have a detachable front that is part wire and part wood, to give shelter in bad weather. Outdoor hutches should be carefully placed and it may be a good idea to change their position according to the season; in the cold months the hutches should be out of the wind and catch all the sun there is, and in warm weather they will be too

SYSTEMS OF KEEPING RABBITS

System	Housing	Feeding	Other Requirements	Problems
Colony	Fenced on grassland with shelters. Fencing needs to be chain-link, buried 6" underground	Grass and weeds they find for themselves in summer. A little concentrate or bran mash for fattening. A little hay should be available.	Water should be supplied. Nest boxes for breeding does.	Expensive to fence Fighting; predators; Escapes. Summer only except in mild climates Babies may get chilled
Morant Hutch Folded on Grass	Wooden ark with run, which is moved to fresh grass every day. Sleeping portion covered in.	Mash or grain to supplement grass, as needed. Hay in wet weather.	Water Nest boxes Wire mesh floor for run to prevent digging. Litter for sleeping part.	Grass may become sour if used too often. — lime after grazing. Summer only except in mild climates.
Outdoor Hutch	Wooden weather-proof hutch made of tongued and grooved board, clad with felt. Good eaves, overhang, part of front solid for shelter.	Wild and garden greens. Hay. Bran mash from scraps — portion size of tangerine for adult. Roots in winter. Feed to appetite — remove uneaten food.	Water drinkers Food trough Nest box. Litter for hutch — straw, shavings, saw-dust, hay cleaning implements. *Claws* need to be clipped every six weeks or so when rabbits in hutch.	Wind and rain in winter — try to avoid draughts. Shorter breeding season. Rats. Dogs and cats.
Indoor Hutch	Wooden or wire cages in a shed. Must be light and airy, with plenty of room for air circulation	Rather less food may be needed. Do not let rabbits get too fat. Try to provide variety since rabbit cannot choose.	Good lighting and ventilation. Other requirements as above. *Claws* clipped. Can use extra artificial light in winter	Snuffles if ventilation not right. Disease in general can spread more easily. No sunshine or fresh air.

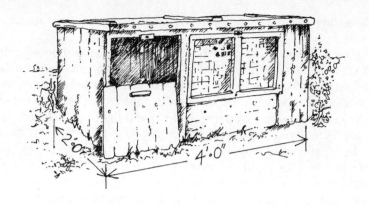

hot in direct sun and may need a little shade. Outdoor hutches are still on sale and if you study the catalogues, this will give you an idea of what to buy or make.

Commercial breeders use wire cages now, but they are not so good as wooden ones and are certainly unsuitable for outdoor rabbits. Two fattening cages for every three doe cages is the normal arrangement. The size varies quite a lot according to the breed of rabbit you decide to keep, since a Flemish Giant is about twice the size of a little Dutch.

65

Breeding

Since they breed more easily in the summer, it will be wise not to start a new rabbit venture in the autumn or you may be disappointed. Starting in the spring will give you some experience before the next winter. Commercial rabbits are kept breeding through the winter by artificial heat and lighting, but backyard rabbits may like their wild brethren decide to take a rest in winter; and since you will expect the breeding does to last much longer than the eighteen months which is the life span of the commercial rabbit, this seems fair enough. Even so, you may expect say six litters a year with about five young each time.

To begin with, it is usual to buy three does and an unrelated buck; so you will need at least four hutches to start with. More room will be needed when the young rabbits are weaned away from the mother, but some backyarders give their does big hutches, and let the youngsters stay with the mother until they are ready for the table. Or you could have one large hutch or pen for the young rabbits; the sexes can be run together because they will be killed before maturity. Rabbits are mature enough for breeding at about five months, but they should be ready for eating at about three months on the backyard system. Commercial rabbits are ready at nine or ten weeks.

The does you buy in may be young ones and if they come to you in plenty of time, they will get used to the place before they

settle down to breeding. When they have reached the proper weight for an adult of that breed, does can be taken to the buck; this is the criterion, rather than the age. If doe and buck do not mate, try again later. Always take the doe to the buck's cage, never the other way round. One month after a successful mating, you may expect a litter.

The doe will need a nest box in her hutch, with some fine bedding in it — good hay is about the best. She will make a nest here, plucking the fur from the underside of her belly to line the nest. She will then look rather moth-eaten, but this is normal procedure, and it makes the teats more accessible for the babies. Rabbits usually have their litters with no assistance and no fuss, but keep an eye on her discreetly because if she is at all frightened she may drop the babies on the hutch floor rather than in the nest and they will soon chill and die.

Baby rabbits are completely defenceless; no sight or hearing, no fur. It is vital that they stay in the nest and keep warm, so be extremely careful how you approach the doe at first since rabbits are nervous creatures and easily upset. If she knows and trusts you, this will make things easier.

At about three weeks, the baby rabbits are furry and lively and they will begin to venture out of the nest and to try solid food. This stage should last for a week or two before they are weaned, so that they can get used to managing with less milk. Make sure that water is accessible to them, but not in so large a container that they could fall in and drown.

The doe can be remated before the litter is weaned. Some commercial breeders save time by remating 3 to 5 days after the birth of the youngsters, but a backyard doe will probably not be worked as hard as this. Other people try the doe with the buck at 21 days after the birth, while others let her go for six weeks before remating. There is a danger that if you leave her for too long, she may be difficult about starting another litter, and may even show a false pregnancy. If she does this, you have to let her get over it and remate when she thinks the rabbits have been born; she will make a nest and so on, just as if they were due.

Feeding

There are advantages for both town and country dwellers when you keep rabbits. Anyone with access to the countryside should be able to gather wild food; the usual kind of vegetation growing on a bank will give them a good meal, with the variety they love. Wild rabbits never eat one food exclusively and a good mixture will ensure a balance of minerals. Consult the list of poisonous plants which appears on page 198 for what to avoid. The sort of handful I gathered for our rabbits would perhaps contain grass, vetches, dandelion, plantain, and so on. It would be clean and free from pollution by traffic or dogs, and probably short and fresh because we used to go back to the same place frequently. If possible it would not be wet, there would certainly be no frost on it when the rabbits saw it, and there would be no soil. It is worth taking trouble with these points to make sure that the food is right for the rabbits; if something does disagree with them, they are very ill indeed because they cannot be sick. Many pet rabbits die because the feed is wrong. With care, you can adjust the diet to the needs of the rabbits; chickweed is good when they are scoured, dandelion has a laxative effect.

Wild greens.

Gardens can provide a great deal of rabbit food. Most of the vegetables we grow for ourselves are acceptable to rabbits and they can deal with the outside leaves of cabbages. Artichoke leaves we do not eat, we grow Jerusalem artichokes for the tubers, but the leaves make a good summer meal for rabbits. They also love sunflower seeds.

Town backyarders have several potential sources of food. The greengrocer is often embarrassed by leaves and vegetables he doesn't want and they make good rabbit food; in areas near sugar beet factories, the beet fall off lorries on corners, right into your hands. Other ideas are bakeries and bread shops, to provide the cereal part of the ration.

These things, together with household scraps, will be the most economical foods for your animals. Conventional rabbit rations consist of pellets costing somewhere over £100 per ton; the rabbits get about 4 oz of these a day, and the ration is doubled for a doe with young. This with hay ad lib and water is the entire diet. Now your rabbits may take in more food than this in order to get the right nutrients, but don't overfeed them. If they leave some, give them less and remove the stale food. If there is always plenty of hay available, in a little wire rack to keep it off the floor, they will not starve. Buy the best hay you can get for the rabbits — or make it yourself from the lawn, or from quiet roadside verges.

Our rabbits got two meals a day, with hay and water there all the time. In the mornings we made them a mash; it was my job and I used left over porridge or crusts, moistened, crumbled and then dried off with bran to make a mash. A portion the size of a tangerine was given to each adult rabbit, in a wooden trough. In the evenings they had green food, or roots in winter. The food was always fresh and clean and we took particular note as to whether it was all cleared up.

Using Rabbits

Killing and dressing rabbits for the table is the biggest problem in some people's minds. This is a pity, because backyard rabbits fed on waste food, are the best kind of independence you can get and they are possible for almost anybody. It seems strange that a handful of nursery tales with rabbits as heroes should put so many people off such a good source of food.

SKINNING

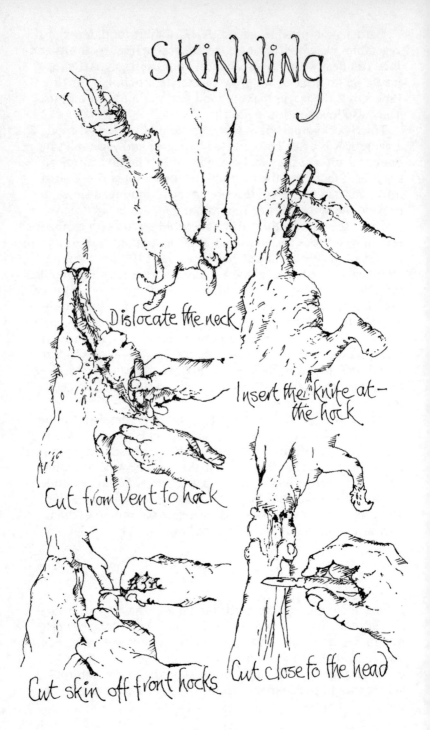

Dislocate the neck

Insert the knife at the hock

Cut from vent to hock

Cut skin off front hocks

Cut close to the head

DRESSING THE CARCASE

Slit the abdomen down the middle using two fingers to guide the knife away from the intestines.

Try to take out everything maybe cutting to release the liver.

Remove gut at the pelvis joint.

One way out is to get someone capable to kill and dress the rabbits for you until you can learn to do it yourself. Most men with a country background will know how to deal with a wild rabbit and the procedure is exactly the same. They are killed by a quick dislocation of the neck, and are skinned and gutted immediately.

Another query in the minds of decent people is whether it is right to keep rabbits in hutches. With plenty of space, I do not think it is at all cruel; wild rabbits spend half their lives underground. At times when you are there to supervise, they can be let out of the hutch to hop around, or they can be given Morant hutches on grass in summer.

There may come a day when your rabbit venture is so successful that there is a surplus of rabbit meat, even after it has been used for bartering. The Commercial Rabbit Association may be able to help if your rabbits are of the kind they require. They collect rabbits alive and pay standard prices, but they really expect pellet fed rabbits of the white breeds. However, if you have a suplus it might be worth investigating. It will of course be illegal to sell dressed rabbit meat from your home.

Rabbit Man

6 bees

Beekeeping is not the sort of enterprise to begin with a book in one hand; it needs knowledge with practical experience. You will need to get among beekeepers and then see the work done, before trying it for yourself. Fortunately, there are beekeepers' associations all over the world and many education authorities run classes in the subject.

Handling bees does not suit everyone; many people have an aversion to insects which they are not able to overcome. This is why it is wise to start handling bees under supervision and to be familiar with them before getting your own hive.

If you can face the prospect of handling bees and getting the occasional sting, they will be a very good enterprise to have, especially for those who live in towns or with little land for larger stock. Bees do very well in towns because the gardens often provide more flowers than can be found in the country-side. A roof top beehive is in nobody's way; the bees' flight path will be above peoples' heads.

Another advantage of bees is that unlike most stock, they do not need attention every day. During the winter they are left alone for much of the time, and in the summer months a visit once every week or ten days is usually enough.

Beekeeping could be an expensive hobby to begin with, but it need not be so. Equipment can be secondhand or home made (see the literature for this, particularly Backyard Beekeeping). Instead of buying pedigree bees you can catch your own! In early summer, there are often swarms about — May is the best time to take one — and householders are often delighted for someone to come and take away the bees from their house or garden.

If a hive of bees is overcrowded or underworked, some of the bees will leave to look for a new home; this is a swarm. They will behave as one body and cluster round the queen. They swarm onto tree trunks or buildings. This summer I heard my brother making arrangements on the telephone. 'Trinity Church at eight, bring gloves and a veil'. It sounded like a wedding, but they were taking a swarm in the nearby town.

The skill in beekeeping lies in keeping the colony up to maximum strength when the honey flow is on, without their actually swarming. Beekeepers try to give the expanding colony more space and see that ventilation is adequate. Young queens are not so inclined to swarm, so some people restock with a new queen each year; young queens also lay more eggs.

Commercial beekeepers have many tricks in order to get all that they can out of the bees, but backyarders can leave the bees to get on with the job without too much interference. Just occasionally they will need a helping hand or a push in the right direction. The main difference between natural beekeepers and the others is that the former only take away the surplus honey. They leave enough behind for the bees' winter food, whereas some would take it all and leave sugar syrup for the bees. This is of course an inferior food, and your bees will be glad of it when they are getting established or in a hard winter, but it need not become their staple food.

The harvest will vary so much according to locality and the season that we cannot really predict results. The honey flow depends on the quantity and nearness of bee forage plants, and this will vary from one year to another, and also according to the weather because they cannot forage in bad weather. A really promising harvest has often been spoiled in our rather cold and windy part of the world when the weather was wild and wet just when some bee plant was ready.

In a good season and in the right area, one could hope for up to about 100 lb of honey from a hive; but there may be none at all another year after the bees have been allowed their winter rations. In a normal year, one hive should supply the family with sweetener for a year, now that so many of us are aware of the dangers of sugar addiction. Those of us who prefer natural food will often use honey instead of sugar, in making wine and preserves and in cooking. A good year may give you a surplus for bartering or to hold over in case of future shortage.

Honey enthusiasts have a long list of its valuable ingredients and uses. It is said to cure stomach troubles and anaemia. It is soothing and makes a very good nightcap dissolved in hot milk. Externally, honey can be used to disinfect and heal wounds — it is a very old wound dressing; and it improves the skin. There is still a little mystery about honey; all its ingredients are not known.

If you have several hives and a lot of honey to spare, you could do as the medieval monks did and make mead — 4 lb of honey to 1 gallon of water, left to ferment.

There is great benefit to neighbouring land and gardens from the bees quite apart from the honey — their function as plant pollinators. White clover, for example, is not fertile without bees. About 80% of fruit tree pollination is done by bees and

75

however small your patch there should be room for a fruit tree or two — in the hedge if need be. If you have no land, gardeners and farmers will be pleased to allow your hives to visit them in summer, or even permanently. There are many crops which do better with bees about.

Bees can go up to two miles in search of nectar and pollen; the nearer the source, the more work they can do, so it pays to have them as near as possible to the flowers. Many people move their hives up to the moors in late summer when the heather is in bloom. Another idea is to grow special plants. They love clover, bee balm *(Melissa)* and aromatic herbs, thyme and rosemary, sweet smelling flowers such as wallflowers and sweet peas, they like dandelion too and that menace of gardens, rosebay willow herb. I have heard it said that poisonous plants such as rhododendrons can pass on their poison in honey, but I have never actually heard of trouble from this source.

Housing

Bees have been kept in very simple hives, but the seemingly more complicated ones make life easier for you. 'Backyard Beekeeping' tells you exactly how to make a hive, as do several MAFF advisory leaflets. This is where practical experience comes

The W.B.C. hive

in; in a beekeeping group you can discuss the merits of the various designs of hive. The bees themselves will finish off the hive; they seal all cracks with a substance called propolis so that it is airtight.

The hive consists of a floor, a deep box and two or three shallow super boxes, an inner cover called a crownboard and a roof. There is also a queen excluder, to keep the queen out of the honey supers. As a compromise you can buy a hive in kit form and assemble it yourself. You will also need a clearing board with a Bee Escape in it, to get the bees out of the supers before removing them.

The hive should be sited to allow for several requirements:-

1. Enough space — not under dense trees or foliage to impede the flight path, which should not cross a road or lane.
2. Water close by to save going to look for it, since bees need to carry water to the hive. A perch such as a partly submerged stone will enable them to get the water without falling in.
3. There should be some shade, some sun and shelter from the wind.
4. Warm conditions — in some countries, a slope to the sun.
5. No damp.
6. Room for you to work near the hive.

Other Requirements

You will also need protective clothing. Gloves, a wide brimmed hat with a veil which is tied with tapes down to the chest, overalls and boots will make you bee-proof and give you confidence until you get used to the bees walking over you.

A hive tool is very useful to prise things apart which have been stuck together with propolis by the bees. It is usually a steel tool with a flat blade at one end and a right angled scraper at the other. A smoker is not cruel; it will help to quieten the bees. It is used to move them about the hive and to subdue them when excited. It is a small can with smouldering paper and hessian inside, which can emit a puff of smoke when needed. It has the effect of making them eat honey, after which they are less likely to sting.

A feeder may be necessary if food for the bees is short. In a difficult time they may need syrup to keep them going; a feeder is usually a container with about four pints of syrup in it, having a perforated lid which is inverted over the feeder hole.

The Bees

There are many varieties, but if you take a swarm you have no choice — except that it won't be a strain which is famous for not swarming! Different varieties are native to different parts of the world, some dark and some lighter, and their temperaments vary as some are more placid than others. Strains of bee have been bred to suit local conditions and this is an argument in favour of buying stock from an experienced local beekeeper.

The native British bees were almost wiped out by Isle of Wight disease in the 1920's but they had been imported into Australia and may survive there in a pure state. American bee keepers have recently been disturbed by reports of fierce South American bees heading north. But — unaided by man it seems they do not cross high mountain ranges.

There are in the colony three different kinds of bee; the queen, the worker and the drone. There will be a few hundred drones and thousands of workers — up to fifty thousand in a healthy colony at midsummer.

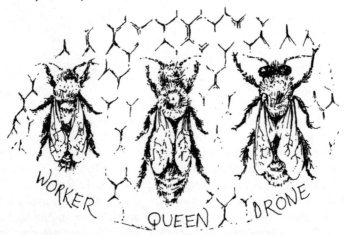

The queen is the only fully developed female, capable of laying eggs. She is the mother of the hive; her job is to lay the eggs in the brood cells. Queens and workers both start as female or fertilised eggs, but the special larger queen cell round the egg ensures that a queen emerges, and when she does, she feeds on royal jelly. Drones hatch from eggs which are unfertilised; beekeepers take away drone cells when they see them since drones are unproductive.

All the larvae actually get the royal jelly, a substance secreted by the glands of nurse bees, for the first few days, but after this the drones and workers get a jelly, honey and pollen mixture fed to them by the nurse bees and only the queen is fed large quantities of jelly. The queen is bigger than the other bees and will live much longer; three or four years perhaps, while the life of a worker is measured in weeks during the busy summer period.

The Worker. Their first job is to feed the younger ones and the queen, and to help with air conditioning the hive by fanning the wings; the temperature and humidity of the hive can be controlled by the bees. Another job is the storing of nectar brought in by the field workers. This nectar is manipulated to get rid of some of its moisture before it is put in the comb and after about three days the honey cell is sealed. A change takes place when the cane sugar produced by plants is converted into up to thirteen different kinds of sugars present in honey, mainly dextrose and levulose.

After about three weeks of hive duties, the worker goes out collecting. Workers have a barbed sting, which sticks into the victim and thus kills the bee. They will usually only sting in defence of the hive, so that danger from bee stings is in the vicinity of the hive. The stings are known to be a good treatment for rheumatism. The best treatment for a bee sting is to remove the barb with a thumbnail without squeezing it and then rub the affected area with a crumpled dock leaf, as you would a nettle sting.

When the worker gets back to the hive, she hands over the nectar to the younger bees; but before this she performs a dance to let the other field workers know the source of her load.

The Drone. Drones come from unfertilised eggs and are reared in larger cells than the worker; because of their greater width, the queen can lay eggs in these large cells without constriction of the abdomen and this means that the eggs can pass into the cells without coming into contact with the semen stored by the queen. This is how the queen is able to lay unfertilised eggs after she has actually been fertilised on the mating flight.

Drones do no work and their only function is to fertilise a queen should there be a need; they will fertilise queens from another hive. They do not survive the winter and are turned out to starve by the workers in the autumn. An interesting point about drones is that since they come from unfertilised eggs their genetic makeup must be identical to that of the queen.

The bees do not actually hibernate in winter, although collection stops. They cluster together for warmth and live on the

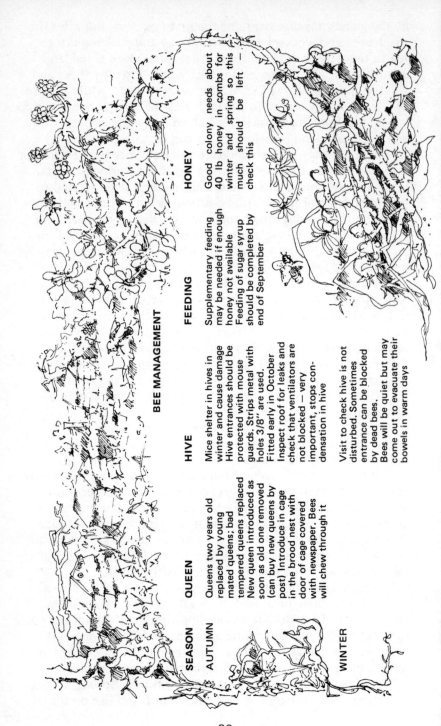

BEE MANAGEMENT

SEASON	QUEEN	HIVE	FEEDING	HONEY
AUTUMN	Queens two years old replaced by young mated queens; bad tempered queens replaced New queen introduced as soon as old one removed (can buy new queens by post) Introduce in cage in the brood nest with door of cage covered with newspaper. Bees will chew through it	Mice shelter in hives in winter and cause damage Hive entrances should be protected with mouse guards. Strips metal with holes 3/8" are used. Fitted early in October Inspect roof for leaks and check that ventilators are not blocked — very important, stops condensation in hive	Supplementary feeding may be needed if enough honey not available Feeding of sugar syrup should be completed by end of September	Good colony needs about 40 lb honey in combs for winter and spring so this much should be left — check this
WINTER		Visit to check hive is not disturbed. Sometimes entrance can be blocked by dead bees. Bees will be quiet but may come out to evacuate their bowels in warm days		

80

SPRING

Check that queen is laying — look at combs in centre which should contain slightly domed worker brood. If a lot of domed drone cappings either queen is laying drones or queen is dead. Colony should be united with one healthy queen

Examine on warm day in April and again in May Remove mouseguards. on second inspection clean up hive and remove unserviceable brood combs. Give extra room as necessary

No feeding should be needed if they were fed properly in Autumn but lift hives gently to check on weight. Feed any hive which seems light. Do it in evening.
Provide a source of water— shallow container in sheltered place out of flight path

Plant flowers the bees like: sweet pea, candytuft, cornflower, stock, panslemon marigolds, mignonette

SUMMER

Queen rearing looked for every week. Queen cells destroyed. If more on next visit it may indicate swarming. Replace queen. Old queen can be put with some bees and food in separate hive

Plenty of room provided for expansion of brood nest and for storage To limit swarming — examine every week Have a spare hive available in case a swarm comes your way

First honey super should be provided when bees begin to occupy outer brood combs. Best of drawn combs When outer combs of this covered add second super. Good colonies will need 3 or 4 shallow supers to hold all their adult bees. Honey taken end of August honey stored in the summer; they will make short flights on warm winter days.

When the queen starts to lay in the spring, the workers begin to forage in the spring flowers for pollen to feed the grubs. By the middle of April the nectar flow has started again. If conditions are right, the population of the hive should have increased by now in order to deal with the increase in work.

Honey

Dealing with the honey can be tricky and this is another thing to study carefully before attempting yourself. An old bee book describes the process thus:

'The honey in the shallow combs (deep comb honey was often sold in the comb) must be extracted as soon as it is removed from the hive, while it is still warm. The extractor is a tin cylinder with an inside cage revolving on a spindle when the handle is turned. The cage will accomodate either two combs, one on either side, or twelve sections, six on either side.

'The first operation in extracting honey is to uncap the combs with a special knife or an ordinary sharp carving knife. A jug, the depth of the knife blade, is filled with hot water and the knife inserted. If a cold knife is used it will spoil the comb. A large meat dish receives the cappings as they are cut off. A comb should be taken in the left hand, held by one lug, the other lug being rested on the meat dish with the comb in a vertical position. The top is then slightly inlined to the right hand in which the knife is held, the cappings will then fall clear. They are removed by cutting upwards from the bottom of the comb with a see-saw movement.

'The combs are then placed in the cage and it is turned slowly. When the honey has been extracted from one side the combs are reversed; extraction is completed by turning more rapidly.

'The honey in the cage is drained off through a tap in the bottom and strained through muslin into jars.

'The wet combs are then given back to the bees to clean down. They carry honey from the wet combs and store it in the brood combs.'

7 goats

There are many advantages of having your own, home-produced milk; our family has not had to buy milk for about twenty years, and we certainly notice the difference when we stray from home! Dairy products supply a great deal of nourishment, especially to a diet without much meat. Think of having fresh milk, butter, cream, cheese and yogurt. Do you really know what fresh milk tastes like, unpasteurised and newly produced and cooled? If you need any more arguments, read that wonderful piece of propaganda, the Backyard Dairy Book.

On the other hand, think hard before you get a goat! They are small creatures and can be kept almost anywhere if you are prepared to work for their happiness; but there are certain basic requirements. Time is the most important. A cow likes bovine company and she will regard you with placid approval, but she won't really want to be petted for much of the time — although the Jerseys do like some attention. The Capra tribe, capricious if nothing else, is different; it will adopt you as a member, with all the responsibilities of leadership.

The Milk

Milk from goats, when sold, comes under the Sale of Goods Act and must be what it says, i.e. no water added or cream removed. It must also bear the name and address of the producer. Cows' milk is hedged about with regulations and licences, which is the result of its bad record in the past, when it has been responsible for passing on such diseases as tubercolosis (almost impossible now, with all the cows regularly tested for it). In fact it will soon be impossible to sell cows' milk in a small way because the last loophole, the Untreated licence which allows farmers to sell their own milk in bottles without heat treatment, will probably be withdrawn soon. But you can sell surplus goats' milk to neighbours, or to a hospital. You could also make it into yogurt and sell it to a health food shop.

This absence of red tape reflects the basic healthiness of the milk of the goat. In a later chapter we will consider milk and its uses in more detail, but let us now consider the particular properties of goats milk.

The average composition is 3.8% fat, 8.6% snf (this means solids not fat, and is the sum of all the constituents except the fat and the water). The fat and the protein are present in smaller particles than they are in cows milk, which makes them more easily digestible. The milk is mildly laxative, it is high in phosphates and Vitamin B. It is usually safe for those allergic to cows milk. In most cases where children have an allergic reaction to cows milk, they can be put on to goats milk with a resulting improvement. This is thought to be partly because it is so quickly digested that the substances in it have no time to cause an allergic reaction before they are broken down into simpler substances. These properties make goats milk particularly useful in medical treatment of ulcers, liver disease and some skin complaints.

The small fat globules make goats milk suitable for cheese making because it is easier to ensure that the fat is held in the curd and does not escape into the whey, leaving a chalky cheese. But the cream is slow to rise and so is difficult to skim off by hand for butter making. The alternative is to churn the whole milk, as they do in Wales, or to separate with a mechanical separator.

Housing

The goat will need a roomy shed, at least 5 ft by 7 ft preferably with a platform for skipping about on and sleeping. An empty hen house will make a goat shed, but keep the goat out of draughts; wild goats choose their sleeping places with great care and though hardy would never bed down in a draught. They have little hair to keep them warm.

84

Take a look at other goat keepers' accomodation and then choose your system according to the space available. The table of systems gives a brief account of the various ways in which people can keep goats happy. Loss of liberty just means that you have to spend more time supplying the food and exercise she would find for herself under natural conditions.

If there is little room for the goat to graze, perhaps she could go out in a yard in front of the shed; she would at least get fresh air there, and some exercise. A goat accustomed to tethering can be used to graze small patches, such as lawns. But a tethered goat should never be left too long alone; you will need to keep an eye on her. A shower of rain upsets a tethered goat'— and no wonder! Or she may wrap the tether round a tree and end up in a twist. If there is someone at home all the time, tethering a placid goat is a good way of using up odd corners and grazing off rough patches. the goat can help with the gardening. But to go out to work and leave a goat tethered is bad management.

In one of their booklets, the British Goat Society give advice to those who must combine goat keeping with a job. They sug-

SYSTEMS OF GOAT KEEPING

System	Housing	Feeding
Outdoor Extensive Grazing only likely in hill areas or woodland	Dry draught proof shed for sleeping, but not warmer than outside air, to encourage thick growth of winter coat. Deep bedding will cut draughts — bracken (dried) could be used	Will find and consume a great deal of roughage, especially in winter. Give access to hay at night. Will be fed concentrates when giving milk
Strip Grazing or Paddocks rotational grazing of good grassland Likely on grass holding with acre or so	If goats have a good growth of 'undercoat' light rain will not stop grazing. Adequate shed should be within reach so they can choose when to shelter. Can be shut up at night and safer	This system gives them fresh grass every day; a good thing with goats as they will not eat what they have trampled on. Kale can be strip grazed in winter. Keep concentrates in dry closed bins. Keep hay in airy shed — Dutch barn is best
Tethering combined with shed BUT NOT for young goats; they need to develop first Likely where space limited	Good roomy goat shed for times when not tethered. Need platform for play and exercise not possible while tethered. Could tether in evening during hot weather	Move tether frequently to provide fresh grazing Try to include bushes and branches within range. Allow plenty of hay to make up for any shortage of feed
Yarding Good shed with concrete yard surrounded by fence or wall. Likely where no land at all	Shed and yard as large as possible. Must have goat-proof doors. Fence or wall high enough for looking out, and holding them in concrete yard	All food is provided by you: be sure of good variety, the secret of good goat feeding. The garden should contribute greenfood Grass can be collected from roadside
Indoors likely in middle of town	Roomy pen with plenty of light and air. Platform for sleeping	Tie up branches for the goats to reach up for. Try for greengrocers' leftovers and neighbours scraps of bread

86

Other Requirements

Salt licks. Goats like a lot of salt. Will need frequent inspection and handling to stay tame. Probably not worth it for backyard goats; better under more intensive system
Access to water

Electric fencing probably best. Three wires at 15, 27 and 40 inches above ground level. Effective but not permanent. Salt licks should be available

Wide collar; tether-chain, swivel, stake and harness
Need shade in sunny weather
Access to water and salt lick. Exercise such as walks.

Take for walks where possible for exercise as well as food
Salt licks needed, also seaweed mineral supplement

Take for browsing walks if possible. You *can* exercise in streets; no different to taking out a dog

Problems

Can easily revert to wild. Some people 'herd' them all day. It can also be difficult to stop goats accustomed to freedom from trespassing in gardens

Breaking out of paddocks is a problem unless fencing is very efficient. It may also be hard to allow access to shelter from each paddock.
Kale will 'short' fence unless it is cut back from it

Flies, rain, cold, wind all problems
Time consuming — you need to check frequently that all is well

Boredom chief problem where goats have no freedom. Variety of food helps

More artificial life for goat but if you can spend time with her she won't mind too much. The less freedom a goat has, the less freedom you have too gest a sensible timetable, with exercise in the early evening — no doubt for you as well as the goat!

Getting Started

It will pay to go to a reputable breeder to buy a good goat. This will not be cheap, but think of it as an investment; she and her descendants could last you a lifetime. To get your eye in, study the prize winning goats at shows. Whichever breed you choose, try to get a good specimen whether pedigree or not. There will be hidden advantages, such as advice and after sales service from a breeder with a professional interest in goats. The chart gives a list of the various breeds.

It may be wiser to buy a young female in kid than a goat in milk. The change of food, surroundings and company that she gets when you buy her constitutes a real shock to any animal. This could easily affect the milk supply and lead to a disappointing lactation; which may be one reason for many people's trying goats for just a short time. A young, in-kid goat will have time to settle down before she is called upon to produce milk.

Whichever way you start, it will be wise to learn to milk before you are called upon to do it. Learners should try on someone else's placid goat or cow, under supervision. If you leave milk in the udder, your adviser can then strip it out. It is very important to get all the milk out of the udder because if some is left, this will encourage the milk flow to decrease. It will be very useful to have some milking practice behind you if your goat is kidding for the first time and is not used to her udder being handled.

Feeding

The British Goat Society booklet on goat feeding is a good guide. It suggests that there are two main problems to keeping goats in a small area, namely providing enough minerals and preventing boredom. The diet can help both of these. Proper minerals will be fed in a varied diet, and plenty of difficult things that take a long time to chew will provide the cooped-up goat with occupation.

Although goats are ruminants like cattle and sheep, that is they digest fibre by regurgitating it into the mouth and chewing the cud, they are not always steady grazers. They also browse,

BREEDS OF GOAT

Breed	Advantages	Disadvantages
Saanen White	Quiet disposition — easy to handle High yielding Good butterfat % in milk 4% Hardy White coat makes easy to see at a distance	Short legs, so udders can get damaged in rough conditions i.e. moorland etc. White coat shows stains. Get skin cancer in sunny countries
British Saanen White	Bigger than Saanen, with longer legs Better on good land Well behaved quiet goat	Rather baggy udder White coat stains
Toggenburg Small, brown and white	Affectionate Good udder Eats less food than bigger breeds Good backyarder, but also good on rough grazing	Very active — may be a nuisance Butterfat rather low Milk yield low
British Toggenburg big, brown and white	Big goat with high yield Great pet Long legs, good on rough land	Excitable — high metabolic rate. Can be difficult to keep in. Poor udder. Rather low butterfat
British Alpine black and white	Big goat, high yield 4% butterfat. Good all-rounder Good for extensive grazing	Independent,character, not such a good pet. (may be a good thing if you are busy). Black coat attracts flies. Can have baggy udder
Anglo-Nubian Lop ears and Roman nose Eastern origin	More flesh (if you want meat!) Good yield Good quality milk, butterfat 5% and high solids not fat Economical — good capacity for roughage	May bleat a lot, which is tiresome

GOAT FEEDING QUANTITIES

from the chart published by the British Goat Society in leaflet 'Goat Feeding'

Food	Kids 1–6 months	Kids 6–12 months	Goatlings	Milkers
Milk	3½—4 pints daily	—	—	—
Concentrates	Up to ½lb daily	½lb—1½lb daily	Summer: ½lb if grazing, 1lb if not. When mated increase to 2½lb—3lb at kidding	Maintenance — 1lb Production — Up to 4lb per gallon of milk. In kid and dry — increase from 2lb to 4lb during last 2 months of pregnancy
Hay	Ad lib	Ad lib	Ad lib	Ad lib, probably 5lb per day
Roots	A little when available	Up to 1lb daily	2lb	Up to 5lb daily
Kale	A few leaves	Up to 2lb daily	Up to 4lb	Up to 6lb (less if cabbage)
Grass and grazing	Grazing or handful of grass	Grazing or up to 2lb cut grass	Grazing or up to 4lb cut grass	Grazing or up to 20lb cut grass
Minerals and Vitamins	Churn Brand (High phosphate) mineral supplement. Cobalt lick	As for 1—6 months	As for kids	1 tsp Churn Brand high phosphate mineral. Cobalt lick. Vitamin D before kidding
Beet Pulp	Small handful when soaked	Increase by handful up to 6oz dry weight	6oz dry weight	6oz dry weight

pick here and there amongst bushes and undergrowth — prickles are no deterrent — and they need a good rough diet to keep them warm. The digestion of the chewed up fibre in the rumen by the bacteria living there is a source of body heat. The goat has a relatively bigger rumen capacity than a cow and can live on poorer land; it is also more capable of keeping warm — hence the thin skin.

This browsing habit combines with grazing; goats eat quite a lot of grass and if it is not possible to let them out much, grass can be cut for them and fed in a rack. But they are fastidious and will never eat anything soiled, so it is useless to expect them to graze after other animals.

With a lot of rough grazing available, goats can roam free, and these sort of goats with access to plenty of roughage are as tough as wild goats. (They can apparently easily revert to the wild; see 'White Goats and Black Bees' by Donald Grant.) They will make milk on land that would not support a cow. But many back-yarders have a rather more coddled type of animal, used to a diet of some roughage, plus cabbage leaves and garden waste, and also concentrates — cereal meal plus protein, often the same cake as fed to dairy cows. A highly bred goat kept under intensive conditions, as many of the show goats must be, can be rather delicate, so treat your new goat with great care until you find out just how hardy it is prepared to be. As with any animal it will pay to study the place it comes from and the sort of management it has had. Any changes in this should be made gradually, so as to cause as little stress as possible. It is possible to 'harden off' a rather delicately reared animal by letting it go outside for increasingly longer periods, but this should be started in warm weather. And the goat's own opinion should be taken into consideration; she will certainly offer it.

To get back to feeding; the accompanying chart was produced by the British Goat Society, to whom I am grateful for permission to reproduce it here. It will give you an idea as to the quantities of food your goats will expect. In practice, most roughages, hay, branches etc are fed 'ad lib' which means you offer humbly what you hope will please, and the goat eats as much as it fancies. If she eats it all up, you trudge off in search of more. But concentrated feeds should be strictly limited — follow the chart and you should not go far wrong.

In town, there should be greengrocers leftovers; during the war, milk records were broken by goats in backyards in the East end of London. In a more rural place, the goats can be taken

91

for walks. A leisurely stroll along a leafy lane, without actually pillaging private property, can be most rewarding. But of course it takes time. You are not wasting time, because she is making milk and you are learning the likes and dislikes of your particular goat.

I always found it best to feed the goat the concentrate part of the ration at milking time, to distract the goat from the task and give it something to look forward to. Some disagree and feed it at another time. Ordinary goats should not need more than 2½ to 3 lb per day; it is simple to buy cow cake and leave it at that, but perhaps more interesting to mix your own ration. A good one to makeup would be equal parts of bran, crushed oats, flaked maize and linseed cake. An alternative would be; equal parts of flaked maize, rolled oats, dry sugar beet pulp, decorticated groundnut cake. By tradition, the concentrate feed is supposed to pay for production, ie milk, replacing the nutrients that went into it. The rest of the ration is maintenance, to keep body and soul together. A convenient theory, but the animals don't know about that.

Goats are very particular, and all the buckets and feed troughs should be kept very clean. They should always have plenty of clean water, even when on a tether. Attention to detail and observance of the goats' needs can be really interesting and can add a great deal to life; but probably more than any other domestic animal, goats demand individual attention.

Breeding

The breeding season, during which females can be taken to the male, is September to February. Female kids are usually kept until about 18 months old before breeding from them, to give them time to grow. They will breed earlier, but they may not do so well.

When ready to breed, goats are in season — calling, wagging the tail and great excitment are all symptoms. This happens every twenty one days during the breeding season and may last for up to three days at a time. Once she is safely in kid she will not come into season again, so after a visit to the male she is watched carefully at 21 days again to see if the symptoms are repeated or not.

Any kind of male will do if you are only concerned to get her milking, but if you want to keep a possible female kid for breeding, go to the best male you can find.

Many goats will 'run through' which means that they will carry on milking and will not need to kid again for two years, sometimes even longer. Unless she is giving a great deal of milk it is usual to give the in kid goat a month or two rest before she kids again; so she can be dried off. A second goat will be handy at this stage to keep up your milk supply. During pregnancy apart from this the goats are milked as usual, but care should be taken as time goes on that she is not frightened or hurried or given bad food.

'Cloudburst' is the name given to a false conception, which can occur with goats. When the time comes for kidding, all that appears is the membranes and a lot of fluid, which is very disappointing. The only consolation is that the goat usually starts milking as though she had kidded.

The gestation period is officially 150 days, but it can vary quite a bit.

When the goat kids, don't worry too much. Buckets of water should not be left in the pen in case the kid is dropped into one. The loosening of the pelvic bones and the tightness of the udder will warn you that the birth is imminent; the goat will be restless and perhaps a bit scared. Normal presentation is forefeet first, hooves facing down, with the head resting on the legs; but a breech presentation is quite usual. If the legs are bent back

93

they will have to be straightened out — send for the vet if you are inexperienced, but watch what he does and learn so that you can eventually do it yourself.

As with other animals, the mother should lick her offspring and this is the best drying and stimulating treatment it can get. You may have to clear the mouth and nostrils to help it breathe, and if she doesn't lick it, rub it well with an old towel. There may of course be other kids and they could be half an hour behind the first, so don't relax too soon. Then will come the afterbirth, one for each kid, which must be allowed to take its natural course. If you never see it and the goat seems unwell after a few days, call the vet.

Do not feed the goat too heavily for a week or two after kidding; the milk supply will gradually increase, as it would in nature to meet the needs of the growing kid, and the food requirements will increase at the same time.

If one or more of the kids is a male, it must be decided immediately what is to be done with him. They grow too rough and boisterous to be kept as pets. Either harden your heart and chloroform him at birth, or make sure you can sell him for fattening or are tough enough to have him slaughtered. This is particularly difficult with goats.

I know some people who take other backyarders' goats — males — as kids and rear them in a bunch for the freezer. They say the meat is very pleasant. Part of their land is strewn with huge boulders round which grow heather and grass, a goat's paradise.

These are very amiable people, but they manage to eat goat without any compunction. I think the secret must be to run them in a small herd; the goats' bond with them will not be very strong as they have companions of their own kind, and the family will not know the goats individually. They reminded me that hospitals are sometimes very pleased to have goat meat for their special diets. It is almost free from fat and easily digestible.

Even the skin of a newly born goat would make something if you were prepared to cure it; Mackenzie in his excellent book 'Goat Husbandry' suggests how to do it.

As Companions

Because goats are affectionate and love company, they can be useful as companions for other species. The most well known

94

example of this is with racehorses; the flighty creatures often have a goat as a stable companion.

Backyarders can use this idea to advantage in many ways. A sow all on her own will be rather lonely between litters; pigs can be friendly with goats. Joan Shields, a well known goat keeper, had goats which used a sow as a step ladder when they wanted to steal apples from the orchard. Young puppies or lambs can be run with goats. They fit into the backyard very well, and their milk is suitable for any orphans there may be.

The British Goat Society have this to say about kid rearing:

Methods

1. Rearing the kid on its dam. This may seem to be the easiest way, but the kids are not so easy to handle, and are difficult to wean, and so cannot be run with the herd. The goat must be stripped daily (This means getting the rest of the milk, if any, by hand).

2. Bottle Feeding. The kids are generally left with the dam for the first few days to get the colostrum, and then put onto the

bottle. Four feeds a day for the first 4 or 6 weeks, the quantity increasing to a total of not more than 4 pints a day for each kid. After that, 3 bottles a day, then 2 and finally 1 when the kid is 6 to 8 months old. Milk substitutes as sold for lambs and calves can gradually replace the whole milk if desired.

3. Pan feeding. This is inclined to produce pot bellied kids, owing to their rapid consumption of milk. Sucking is better for the kids digestion, being more natural.

Kids will nibble hay and concentrates at a very early age and should have plenty of fresh air and exercise.

Fencing

Goats are extremely difficult animals to fence in or fence out and restraining them needs foresight and cunning.

It really is best to spend money at the outset and fence properly if you intend to let the goats go free outside; you may have to tether them until this is done. They can do so much damage in so little time, browsing in gardens, that you cannot afford to risk escapes. Goats on the loose make for bad relationships with neighbours quicker than anything else; five minutes talk with a goatkeeper will tell you this and it is, unfortunately, the most usual reason for people giving up goats. But it is not impossible to fence them in, and they are worth the trouble and expense. So do it before they start to cause trouble because once they get a taste for wandering, they will be even more difficult to keep in.

The problem is the height to which you must fence because they can leap over obstacles and run up walls. The best solution I have seen is to fence the garden with chain link fencing on stout posts up to about eight feet. On this holding, the goat field is walled with a stone wall and chicken wire has been added on top of this. Two layers of pig netting, one on top of the other, would be effective, but the posts must be high and strong.

Permanent goat fencing looks rather repressive and electric fencing can be a good alternative for temporary grazing, but this must be arranged with care. To be safe you will need three strands of electric wire at strategic intervals with a really good battery or mains fencer unit.

8 pigs

The easiest way to start pig keeping is to buy two 'weaner' pigs and fatten them for pork or bacon. They should be easy to find from either a backyarder or even a pig farm, because the specialisation in the industry means that many farmers keep sows and produce weaned piglets to sell for other people to fatten. So the weaner is a recognised commodity and you are fairly certain of getting a fair deal; in other words, it is better to buy at this stage than a pig in the middle of its breeding life.

The weaners will come to you at about eight or nine weeks old and they will be ready for pork in six months or so, perhaps a little longer on cheaper food. A month or two more will take them up to bacon weight. They should be easy to care for and a good introduction to pig keeping; the ingenuity lies in finding cheap nourishing food for them to cut the cost of the meat. Food costs amount to about 80% of this.

The breed of pig you choose will most likely depend on what is available locally; see the chart of pig breeds for details. The older breeds, often coloured, are a better bet for backyarding because they will probably have escaped modern intensive management. Pigs have been developed to grow very fast indeed, but on high powered food, often laced with antibiotics to keep down disease, and in a controlled environment with no contact with the great outdoors. It is not realistic to expect such highly bred pigs to rough it. If old breeds are hard to come by you could buy a couple of hybrids and try to naturalise them slowly!

First of all, check the chapter on regulations and see that

nothing in your tenancy agreement, covenant, conditions of purchase or bye-laws will prohibit the keeping of pigs. When we started in a semi-detached suburban house with our back-yarding, the landlord stipulated no pigs, so we had to wait until we got a smallholding for pigs. On the other hand, some allotments run a pig club, so it may be possible to keep a couple on an allotment.

It may be a bit tough getting started. Some people like pigs and some are violently against them. If you let it be known that you intend to get some, the neighbours may be alarmed. They will quote smell, pollution, flies and the lowering of the neighbourhood tone.

True, pigs do smell — factory farms are sometimes very smelly although they are very clean inside. The stored effluent causes the trouble. But two small weaner pigs, into the freezer before they reach maturity, do not produce much odour at all. If you can confine the pigs to where they ought to be, your weaners will not be a problem. A gift of manure for the garden will often sweeten the opposition!

Housing

If you have an empty wash house, a deserted garage or the old fashioned pig pen at the bottom of the garden, you are lucky. A high building can be made warmer by fitting some hardboard about 4 ft above the pigs' bed. A pen can be made in a big building by making solid but moveable hurdles, tied together

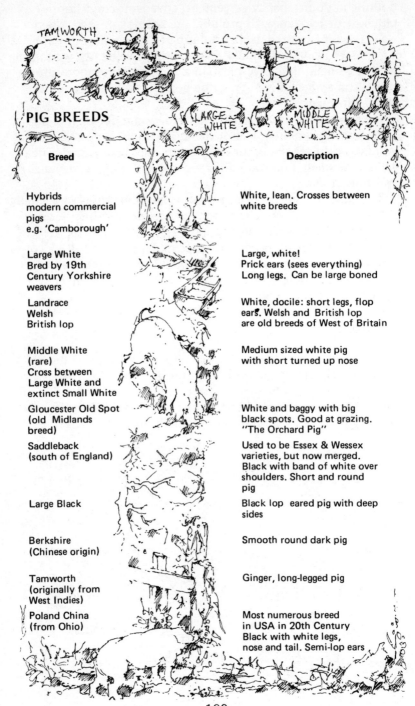

PIG BREEDS

Breed	Description
Hybrids modern commercial pigs e.g. 'Camborough'	White, lean. Crosses between white breeds
Large White Bred by 19th Century Yorkshire weavers	Large, white! Prick ears (sees everything) Long legs. Can be large boned
Landrace Welsh British lop	White, docile: short legs, flop ears. Welsh and British lop are old breeds of West of Britain
Middle White (rare) Cross between Large White and extinct Small White	Medium sized white pig with short turned up nose
Gloucester Old Spot (old Midlands breed)	White and baggy with big black spots. Good at grazing. "The Orchard Pig"
Saddleback (south of England)	Used to be Essex & Wessex varieties, but now merged. Black with band of white over shoulders. Short and round pig
Large Black	Black lop eared pig with deep sides
Berkshire (Chinese origin)	Smooth round dark pig
Tamworth (originally from West Indies)	Ginger, long-legged pig
Poland China (from Ohio)	Most numerous breed in USA in 20th Century Black with white legs, nose and tail. Semi-lop ears

100

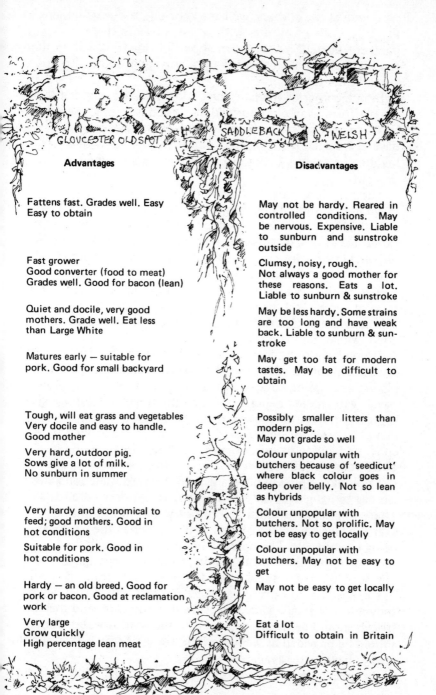

Advantages	Disadvantages
Fattens fast. Grades well. Easy Easy to obtain	May not be hardy. Reared in controlled conditions. May be nervous. Expensive. Liable to sunburn and sunstroke outside
Fast grower Good converter (food to meat) Grades well. Good for bacon (lean)	Clumsy, noisy, rough. Not always a good mother for these reasons. Eats a lot. Liable to sunburn & sunstroke
Quiet and docile, very good mothers. Grade well. Eat less than Large White	May be less hardy. Some strains are too long and have weak back. Liable to sunburn & sunstroke
Matures early — suitable for pork. Good for small backyard	May get too fat for modern tastes. May be difficult to obtain
Tough, will eat grass and vegetables Very docile and easy to handle. Good mother	Possibly smaller litters than modern pigs. May not grade so well
Very hard, outdoor pig. Sows give a lot of milk. No sunburn in summer	Colour unpopular with butchers because of 'seedicut' where black colour goes in deep over belly. Not so lean as hybrids
Very hardy and economical to feed; good mothers. Good in hot conditions	Colour unpopular with butchers. Not so prolific. May not be easy to get locally
Suitable for pork. Good in hot conditions	Colour unpopular with butchers. May not be easy to get
Hardy — an old breed. Good for pork or bacon. Good at reclamation work	May not be easy to get locally
Very large Grow quickly High percentage lean meat	Eat a lot Difficult to obtain in Britain

This table by kind permission of Practical Self Sufficiency with wire at the corners; we have kept pigs for years in pens like this, made of hurdles with wooden frames clad with second hand corrugated sheets. A pen about 8 ft by 10 ft will be almost too big, but it will fit them as they grow and would also house a sow in the future.

In one corner of the pen, the darkest, make a wooden sleeping platform — an old door will do. Wood is so much warmer to lie on than concrete and pigs feel the cold, especially when they have just parted from a large litter. Plenty of straw or other bedding on the floor will also make them happy and cut down draughts.

The pen floor must be of concrete, otherwise smells will be inevitable. It should ideally slope to an open drain which will take the liquid effluent outside the building. You need not lay expensive drains, because the liquid can be caught outside in an old bucket or paint tin and will be a good compost activator or liquid manure. A good slope will let it run clear of the straw; pigs go to the light to dung, so the bed in the dark corner should be kept clean. And the straw will give them something to root in if they are not allowed outside. Barley or oat straw is very good, wheat is a little stiff.

If you have to build a special pig house, the traditional cottage sty cannot be bettered. This again is a compromise for the pig between being outside, which is not possible if you only have a garden, and being indoors all the time. The cottage sty is pig sized; the hut is low — it can be as low as 4 ft but 5 ft makes it easier for you to get in. In front of the hut is a yard, concreted and with a wall or good strong hurdles round it. The hut can be about 8 ft by 8 ft and the yard may be any size you like. You will need a gate in the yard wall, and a hessian bag to hang over the hut doorway to cut down draughts. There will be no need for a door but you could fit one so that the pigs could be fastened in the sty when the yard was being cleaned out. The opening in either case should be to one side rather than in the middle, to cut draught again. The pigs should dung in the yard and keep the hut clean; they will like deep straw to bury themselves in, especially at first when they are very small. They can also eat in the yard, and the trough should be firmly anchored to the wall. A water trough should be provided as well as a food trough because they should always have water available.

Pig housing needs of course to be stronger than that for poul-

try and rabbits. The ideal building is probably a cottage sty of
old bricks or recycled concrete blocks. A snug wooden hut can
be made pig proof; and we have made the yard walls from
wooden frames clad with corrugated sheets. A wooden roof will
need to be clad with roofing felt to keep out the rain. If you
have not kept pigs before, take a look at somebody else's before you start so that you have an idea of what to expect.
Weaners weigh about 50 lbs or so, and pork pigs will be anything from 120 lbs upwards. A bacon pig will weigh over 200
lbs and a sow will be about 300 lbs; but these weights will
mean more when you have seen a few pigs and got used to the
sizes.

Pigs on a holding with some land can be very useful indeed.
You can get away with less housing; a straw bale hut, protected
with wire netting so they can't eat the straw, may be all you

SYSTEMS OF PIG KEEPING

System	Housing	Feeding	Other Requirements	Problems
Outdoor Grazing	Straw bale hut or Roadnight corrugated shelter. Can reclaim land with this system	Where grass is plentiful, feed once a day with meal or other concentrated food, according to condition of pigs. Plenty of fresh water available	Deep bedding in shelter to give comfortable sleeping place. Good fences — electric or pig wire	Worm larvae build up if same plot used too long. 'Poaching' in wet weather High piglet mortality Will spoil good grassland *unless* rung to prevent digging
Limited Grazing	Ark or moveable house with yard Or pigsty, in at night etc.	According to amount obtained by grazing. Can supplement with swill also	Deep bedding in sty Very good fences needed	Worms and wearing out of pasture where only small area available
Tethering	Moveable house accessible to tether	Allow for grass and feed accordingly. Never assume that grass will be enough	Harness, tether, chain, swivel, stake. Deep bedding in hut	Many. Sows have to be 'broken in' to tethering. Tethers may get tangled. Harness can cause sores Frequent moves essential
Cottage Sty plus Yard	Sty can be wood, brick or blocks 8' x 8' x 5' to eaves with concreted yard, sloping to drain	Swill or garden waste for part of ration. Potatoes comfrey etc. can be grown for pigs. Give a turf for minerals	Good troughs for food and water, fastened down in yard. Deep bedding in huts. Barrow and shovel for mucking out	Pig dependent on you for all food, including minerals. Restricted pigs are not quite so fit
Indoor Pig Sty	Strong pen with concrete floor, sloping to open drain. Wood sleeping platform Good windows to open in summer. Underdraw roof in winter	Weeds and grass given, turf for minerals Lehmann system of feeding	Bedding in sleeping part Infra red light and creep area for breeding sow Piglets need dosing with iron at about 4 days	More artificial: let out exercise if possible. May get lameness and stiffness in sows

need to fatten your weaners on — and some of their food can come from grass. But the real value of the pig outdoors is in its snout. Pigs can reclaim land, eat off the vegetation, uproot the bushes and dig it all over, manuring as they go. The best way to run pigs is probably like this, but with an indoor sty to fall

back on in wet weather. Outdoor pigs in a sea of mud are not happy and will not grow well. If you have limited space but would like them to have a natural life, let your pigs out at intervals, such as weekends when you can be there to keep an eye on them. If you don't want them to dig up the grass, rings in their noses will stop them.

Starting off at about 50 lbs weight, the weaner pigs will grow by as much as a pound of weight a day, given a warm bed and the right food. Commercial pig farmers reckon that fatteners increase by about 10 lbs per week, but speed is not all with backyard pigs, so don't be disappointed if your pigs seem rather slow growing. They will inevitably stand still for a while when they first arrive, until they get used to the new surroundings; and even them, the food you give them will be cheaper than conventional fattener rations which cost over £100 per ton.

If you want to monitor their progress, get a 'weighband' and measure the girth from time to time. This will tell you the weight approximately. Feed them on whatever they are used to at first, and make changes gradually.

Feeding

Weaners should be eating about 2½ lb of meal a day; they may have been reared on pig grower pellets which contain all kinds of additives. If you were to buy all their food you could change them onto 'sow' meal, which is barley meal balanced with protein. Since it is sold for breeding sows it has no antibiotics and is a more natural food. If you were to keep on with this, you would increase the ration by ¾ lb each week until they were eating about 5 lb per day (Divide into two and feed half at each feed for all these amounts.). Two pigs will eat better than one and will be company for each other and no more work, which is why I suggest a couple of weaners.

Pig food is most of the expense — once the sty is done it should last for years. They will no doubt fatten on pig meal, but the cost can be worked out before you start and it is expensive. I would also think that the pigs would prefer a bit of variety. So it is worth some thought and effort to find a source of cheaper pig food. The digestion of the pig is similar to our own and it cannot digest large amounts of grass and weeds, although it will appreciate these things and will turn garden waste into manure. But to grow well the pig needs protein and carbohydrate without too much bulk. Therein lies the problem.

Usual alternatives to bought in pig food are crops grown in the garden such as potatoes; household scraps; other peoples waste scraps; and factory waste from food processing factories such as bakeries and dairies. Farm waste such as milk pipeline washings, waste milk, potato chats (small ones that pass through the riddle when the potatoes are graded for size) may be available if you live in a rural area.

Waste food or swill is the traditional pig food and in the war this fed many a backyard pig. Unfortunately, since those days, the bones of imported meat have been found to carry disease and so the authorities look upon swill gatherers with grave suspicion. Pigs love swill, for food suitable for us is good for them with the exception of citrus fruits. To make meat out of throwaway food is very satisfying.

The rules about swill are quite strict, and you need a licence and proper premises. The waste food has to be boiled for an

Quantities of barley meal (or its equivalent) to be fed to pigs according to live weight.

Live weight (lbs)		lbs barley meal or equivalent per day
20	ad lib creep feed	
30	ad lib creep feed	
40	ad lib creep feed	
50		2.5
60		2.75
70		3.0
80		3.25
90		3.5
100		3.75
110		4.0
120		4.25
Fatteners usually level off here these days		
130		4.5
140		4.75
160		5.0
Even traditionalists level off here		

Remember that pigs vary and always feed according to condition. I will repeat that the amounts given above are to be fed *per day* —usually half at each feed.

---

hour and the swill setup would be out of the question in an suburban garden. For backyarders with more land it might be worthwhile, especially if a few families combined their resources. and set up a communal boiler. Our family boiled swill for years and produced pigs for sale with it, as part of our plan to earn money to buy more land. We found it hard work, but not too unpleasant and very nice for the pigs. They got an exceedingly varied diet because we collected from greengrocers' and fish and chip shops.

Without boiling it is possible to feed scraps provided the pigs are on a vegetarian diet. Rigorously exclude any meat or eggs from the food — and this includes bakery waste if sausage meat is in it — do this whether you think you are being watched or not, all the time. It will then be lawful to feed your swill without cooking it first. You will then be quite sure you are not helping to spread disease.

107

It may take trouble to arrange at first, but it is a worthwhile project. If anyone saves scraps for you, give them a clean bin labelled *Pig Food. No Meat or Eggs,*and collect it frequently while it is fresh.

Another problem with swill is how much to give them. It is difficult to know the precise value of the stuff. We have to make an intelligent guess and feed to appetite. The list of barley meal equivalents may give you an idea of how much food your pigs ought to have. There is also the Lehmann System; he suggests that your weaner is kept to its original meal allowance of 2½ lb per day, and as it grows the rest of the ration is made up of an ever increasing amount of the bulky foods; grass, roots, swill. The section on growing crops for animals may give you some ideas.

Sow feeding depends on whether she is feeding a litter or not. As a general rule, a sow without a litter should get about 5 or 6 lb of meal a day or their equivalent. She will be producing another litter (one hopes) so this will take some of her food; so feed according to her condition. If she looks thin, step up her allowance. And if she gets too fat, cut it down so as to make things easier for her at farrowing.

When she has her litter the food should be increased gradually until she gets about twice as much as she did before — say about 6 lb per *feed* or 12 lb per day. The rule is 3 lb for the sow and 1 lb of meal for each piglet — so 12 lb meal supposes that she has a litter of 9. But there is a limit to what she can deal with, so a pig with a large litter may have to live off her back for a while and will go thinner. Here again, try to feed according to condition and if a sow gets thin, feed her up again as soon as you can. This may mean forsaking the bulky foods and giving her more meal for a while.

Woodland Foods. Nuts are traditional pig food — it has been suggested that the pig is not a native of the naturally treeless parts of the earth and could not even be kept there until modern times.

Pigs of 100 lb weight and over can safely be fed acorns, horse chestnuts and beech mast where they are easily available. It is suggested by the nutritionists who advised farmers during the first world war that before acorns are used, they should be spread out in thin layers to dry, as when dry they are more palatable and less astringent. They have a high starch content, but should only be fed at a rate of 1 to 2 lb daily along with a more

laxative food. This means that acorns can replace about a third of the meal ration, a good saving.

Horse chestnuts can be dried, husked and ground to make an even richer feed than acorns; 1 lb of chestnut meal equals 1 lb 1 oz of barley meal, or 1 lb 4 oz oats or 1 lb 8 oz bran. But chestnut meal tastes bitter, so the pigs will tend to reject it unless it is mixed up with something tastier. Up to about 1¼ lb per head can be fed per day.

The kernel of beech mast is fatty and can be fed in small quantities.

Barley Meal Equivalents

The following food quantities are roughly equivalent to 1lb of barley meal in feeding value.

Wheat meal	1 lb
Ground Oats	1.1lb
Crushed Oats	1.45 lb
Potatoes (cooked)	4 lb
Dried Potato slices	1 lb
Potato peelings (steamed)	5 lb
Sugar beet (grated)	3.5 lb
Molassed Beet Pulp	5 lb
Fodder Beet	5 lb
Fodder Beet	5 lb
Whey	1¼ galls
Swill (good)	3 lb
Grass	10 lb
Kale	7.5 lb

Breeding Pigs

Beginners with pigs are wise to take it easy at first, to try fattening a couple of weaners and to do costings. If you find that all goes well and the pigs are fitting into the rest of the holding, as pigs often will, the day may come when you want to breed your own.

One way is to buy in a pair of weaners, making sure that one of them is a good looking female with at least twelve teats. At about eight or nine months, when her companion is ready for bacon, this young gilt will be ready to take to the boar. Four months later she will farrow.

There are advantages to this proceeding. It is a relief to be able to look your pig in the eye and talk to it! The pig in turn will know and trust you, and a good relationship is invaluable with a breeding animal. And also, having been reared on the premises, she will have acquired immunity to whatever bugs are present; this she will pass on to the young ones and it is a great asset.

There are snags of course; the behaviour of a gilt with her first litter is unpredictable, whereas most sows know what to expect and get on with the job. I have seen gilts go into a fit of hysterics at the first sight of a piglet crawling towards the teat. The aversion seems to be sight more than anything; they will sometimes let the piglets suckle provided that they can't see them.

Piglet lives can be lost in this way because frightened gilts can savage their litters, and piglets are valuable, so this must be prevented.

If you call an old pigman, he will suggest a bottle of stout; but a frightened pig may not drink it. The vet will give her a tranquilliser, which will solve the problem because when she comes round she will accept their presence — I have done this many times because it seems to me a justified use of drugs. Left to yourself, with no drugs and no beer, the only thing is patience. You will have to take them away from her as they arrive and put them in a safe warm place — a box or a barrel, under a lamp. Then try introducing them to her teats by stealth after you have soothed the pig and got her to lie down by stroking her belly — this can be practiced before the event! If you stop her damaging them, before long she may settle down to feed them and even if you have to take them away and feed them every hour under supervision, patience may be rewarded as she gets used to the idea. But remember the vet

has a good solution, his fee will be more than justified if he saves a whole litter.

An older sow with experience will be easier for your first litter, but good sows are sometimes very hard to come by. In general I would not recommend buying animals in markets; if you hear of somebody going out of pigs and selling them all, try to find out why they have become disillusioned before you buy! Treat a new sow gently and don't expect her to be friendly all at once, or to brave the weather if she has been indoors all her life.

Whichever way you acquire your pregnant female, the day of farrowing will arrive. Whichever system of pig keeping you have adopted, a sow needs a clean pen for farrowing, with some clean straw in it. Give her several doses of bran for a few days before she is due. Get an infra red lamp ready if you intend to use one — but be sure the sow cannot reach it, or it will be broken.

Shortly before the birth she will make a nest, milk will appear at the teats and she will be restless. Then she will lie down, trembling a little and sometimes will appear to be in a coma as the piglets are born.

Some say tiptoe away at this point and let nature take its course; with a sensible pig you can do this, but don't go too far — she may need help before the end. A restless sow may trample new piglets and you could rescue them as before and restore them later — but she may be more restless because you are there! Try to observe her without being seen. Odd piglets can be suffocated before struggling free of the membrane and your presence could save one or more. If she strains a long time with no effect, this could mean a blockage; vets and experienced pigmen disinfect hand and arm and gently investigate.

Bringing up the piglets is no problem with a good healthy sow; she will do the job. If the litter is very large or she seems to have little milk, the piglets can be taught to drink from a bowl. Put warm milk in shallow dishes in the creep area and this may save their lives.

If you decide to wean fairly early, the sooner they are on to solid food the better; so creep pellets will be fed from about a week onwards, always accompanied by a dish of water for the piglets to reach. Creep will also help a large litter to survive. Bought in creep feed is very expensive but the sow meal plus a little molassine meal to make it sweet will be a good substitute.

Indoor piglets will need a dose of iron; paste can be put on their tongues or the iron can be injected into the ham, 2 cc for each piglet. The question of castration now arises. Those planning to sell surplus weaners to a fattener will have to castrate the males, but if you can rear them yourself or barter with another backyarder, they can be left unmutilated. Pigs are ready to go before they reach maturity and so castration is not essential; the boar meat from a young animal is not tainted. The practice is left over from the days when a pig took eighteen months to fatten. Castration of pigs is usually surgical, done with a scalpel and sterile blade. It is a job which can easily be learned, but it is not a pleasant one.

9 the house cow

The cow is hardly a garden animal, since even the smallest breed will need about an acre of land; three acres to the cow was the old standard, including of course the hay for winter keep. When farming papers quote modern stocking rates of one cow to the acre they refer to intensive farming with large applications of nitrogen. On the whole, cattle are easier to look after than goats and will give more milk, produce more manure and a calf that may have good future potential. Many people graduate from goats to cows as they get more or better land, except the goat lovers, who merely increase their herd. Cows and goats could fit in together, when the cow is dry the goat can be milking and so on.

What will be needed besides the land? A cowshed is not essential for a house cow because she can be milked outside if necessary and can live outside for much of the time, depending on the climate. On wet heavy land on the other hand, she may have to be kept off the grass in winter. At the very least I would provide my cow with a shelter in which to sleep in the winter. Cows can lie outside; but to produce milk they need some consideration and the best arrangement is probably one where the cow herself can choose whether to stay out or come in.

The conventional cowshed where the cow is milked and sleeps at night in winter has the following measurements:

Manger 3 ft
Standing 5 ft
Dung passage 3 ft
Back walk (raised up from dung passage) the remaining space

Where milk is produced for sale, there are many regulations about cowsheds which can be ignored for the house cow; but if you happen to inherit one of these sheds it will make life easier. The walls will be cement-rendered smooth so that splashes can be cleaned off, the floor will slope to a drain and be easy to clean and so on. Proper ventilation is important for cows where they have to spend some time indoors, but if she only comes in to be milked and sleeps in a shed with the door open, ventilation will not matter. There is usually outlet ventilation in the ridge of the roof and inlets somewhere lower down, at the eaves or lower. Cows kept in a fuggy atmosphere are sometimes less keen on their food and can suffer from digestive upsets, strangely enough. Then, when they do go out, they may catch a chill. Cows outdoors a lot of the time grow nice thick winter coats and are tougher.

Methods of cow keeping are in fact as varied as those for goats; in commercial herds there is a lot of variation, from outdoor herds milked in the fields through moveable 'bails' to the other extreme, cows kept in yards all the year round and fed from tractor-loads of green food gathered for them. The more freedom you give to a house cow the easier life will be for you and her, but the more room you will need.

The breed you choose will depend on locality and what is available, and the choice is outlined in the table of breeds. Smaller breeds are very good for house cows, being economical producers. The Jersey is probably the favourite, partly because it is such a beautiful animal. Originally bred on a small island, Jerseys were used to grazing on a tether or in very small fields; they were used to human company and being handled a lot and their modern descendants still show a lot of these characteristics.

During any discussion of Jerseys somebody will say that the bull calves are worthless, being no good for beef. It is true that a Jersey heifer is worth much more than a bull and you will not get very much for a young bull calf; but this is not all the story. Where there is grass enough to keep the calf, he could be kept and would make acceptable beef in the end. Jersey beef is not inferior, but there is less of it — not enough to interest commercial fatteners. The fat is yellow, which happens to be unfashionable at the moment.

From our experience I know that the Jersey can be crossed with a beef breed to give a good beef animal; we have used our Hereford bull on well grown Jersey cows and got a tiger striped

JERSEY

animal that seemed to do better than the Friesian — Hereford crosses, its more conventional companions. Where space is limited and the calf will have to be sold, a beef bull can be used to get a more valuable calf. The beef bull should not be used on heifers but it is safe on cows and the Jersey is famous for its ease of calving. The Aberdeen Angus breed is a smaller beef animal and this can be used on Jersey heifers; otherwise use a Jersey bull.

Buying a cow

There are pitfalls in buying any animal already in production, as we saw with pigs. Never buy a cow in a market! By this I mean the open markets; the various breed societies use the market premises for their own sales from time to time, and the animals sold there will be good ones, though no doubt expensive.

The best place to buy a cow in milk is a dispersal sale. Where the owner has died and there is a good reason for the sale, all the cows will be sold and a middle aged cow, sensible and placid, will be the best buy for a house cow.

A study of cows at agricultural shows will give you an idea of what to look for; looks are not everything and do not always give a clue to production or temperament, but any animal will last longer and be healthier if it seems well put together. The udder should not be too pendulous. Check the teats for size; very small teats are quite common now because most cows are

MILK BREEDS OF CATTLE

Breed	Advantages
Ayshire red and white wedge shaped	Good forager. Hardy, healthy Good milker — well shaped udder Milk useful for cheese. Cheaper than Friesian
British Friesian Large, black and white	Large quantities of milk Bull calves can make good beef. Sell well
Dairy Shorthorn	Was most popular until Friesians took over Adaptable, but best on good land. Bull calves good for beef. Butterfat quite good
Dexter Little black cow, Irish	Smallest breed of cattle: mature cows weigh 650lb. So eat correspondingly less and good for backyarding in small space. Does well on poor grazing. Calves useful for beef, small joints
Guernsey native to Guernsey and Alderney	Live a long time Cream rich yellow, rises quickly High butterfat milk
Jersey numerically the largest breed in the world	Matures early — calves at 2 years old Good butterfat, cream rises fast Tolerates heat, good in tropics Nice temperament, family pet Smaller, eats less. Not too expensive Lives long — 15—20 years. Beautiful
Kerry	Used to be widespread in Ireland so are tough, used to poor land. Live a long time good producer. Butterfat fairly good The 'poor man's cow'
Red Poll	Really dual purpose, very beefy No horns. Quiet
South Devon	Some strains with no horns. Dual purpose milk and beef. Long lived. Good cream production
Welsh Black	Very hardy. Good for exposed situations and poor pasture, and wet climates. Dual purpose: some strains good milkers. Long lived

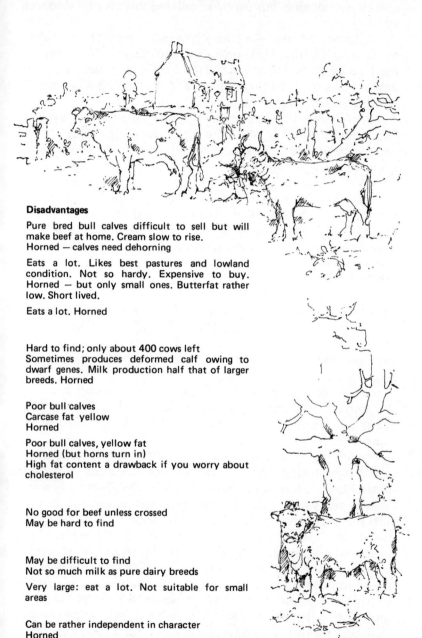

Disadvantages

Pure bred bull calves difficult to sell but will make beef at home. Cream slow to rise.
Horned — calves need dehorning

Eats a lot. Likes best pastures and lowland condition. Not so hardy. Expensive to buy. Horned — but only small ones. Butterfat rather low. Short lived.

Eats a lot. Horned

Hard to find; only about 400 cows left
Sometimes produces deformed calf owing to dwarf genes. Milk production half that of larger breeds. Horned

Poor bull calves
Carcase fat yellow
Horned

Poor bull calves, yellow fat
Horned (but horns turn in)
High fat content a drawback if you worry about cholesterol

No good for beef unless crossed
May be hard to find

May be difficult to find
Not so much milk as pure dairy breeds

Very large: eat a lot. Not suitable for small areas

Can be rather independent in character
Horned

milked by machine, but for hand milking this may be inconvenient.

If you buy a heifer — a young cow either just calved or about to calve for the first time — you will probably find that her teats are small, but they will stretch a little with use. At college where we learned to milk by hand and machine, it was noticeable that the hand milked cows had longer teats.

A heifer would be a good buy in some ways but not in others. She should be healthy, will have a good long life in front of her and can be adapted to your ways. But she will not be used to being milked and patience will be needed! I have taught many heifers to be milked, and found that quiet but firm handling is essential. The Jerseys seem to be the quickest to learn to behave, probably because they are not usually nervous. Heifers may kick a lot at first, through fear. They soon learn that nothing will hurt them, but of course if you are not used to milking, trouble could ensue. Learn on a cow before you tackle a heifer.

TB is not usually a problem now because cows all over the world are regularly tested for it and reactors disposed of. Brucellosis is going the same way, but at the time of writing in Britain not all herds are brucellosis free, so do check that you are buying a cow from a herd which is accredited, which means that it is blood tested for the disease.

Brucellosis is an unpleasant disease, causing contagious abortion in cattle and undulant fever in man. If you are not sure about a prospective cow, the vet would do a blood test with the owners' permission.

The scourge of commercial dairy herds and the most common reason for getting rid of older cows is mastitis, a condition of the udder. Hard, lumpy udders should be avoided. Clinical symptoms are heat and swelling of the udder and clots in the milk, but the clots don't always show up. Various organisms are present in mastitis and are blamed for the disease, but they can sometimes be present without causing the trouble. Stress of various kinds seems to be the predisposing factor and some cows are born more nervous than others. There must be quite a lot of stress in big herds, competition for food and so on, which your little house cow will avoid; and also, you will not be pushing her for maximum yield. Try to get a placid cow with a good appetite and treat her gently. Nervous cows are a great source of stress to their keepers.

Breeding

Sooner or later your cow will dry up; she should have about two months' rest before she calves again, although it will depend on how long she herself is prepared to go on milking. Cows should never be hurried or frightened, and as she approaches calving, gentle treatment is even more necessary. If you have to dry her off, milk her once a day for a week or two to discourage the flow.

As calving time approaches, the udder will fill up in size and the pelvic bones will drop a little; a cow about to calve is restless. She will often paddle with her hind feet. They can usually manage themselves, but keep an eye on her in case anything goes wrong. A cow which has been straining for some time without any results may need help and the vet is the best person to give it.

When the calf is born she should be encouraged to lick it dry and the calf will need to drink fairly soon; it should have had its first feed by the time it is six hours old. The calf can be left with the cow for the first day or two unless she deliberately ignores it, or is frightened of it — which can happen. Left with her for longer, it will be difficult to teach to drink from a bucket and the cow will be much more upset when the parting does come.

Some people try to combine a house cow with a suckler cow; they leave the calf with the cow, and milk out the dregs that the calf leaves for themselves. This is one way to do it, and if you have very limited time and everything goes well, it can be successful. But if you want to make dairy products and to have control over what is going on, the simplest thing is to wean the calf at a few days and feed it on its mothers milk in a bucket; then you will know how much she is giving and how much the calf is drinking. They will be easier to control if they are separate.

Getting the cow in calf again should not be too difficult. Wait until the second heat after she calves, then decide what bull you want to use and ring the local AI centre early in the morning of the day she is in season, and they will inseminate her the same day. Unless you happen to have a near neighbour with a suitable bull, this will be the easiest way to get her in calf. Remember to watch her after 21 days or so to be sure she does not come in season again; if she does she will have to be served again, which the centre will do for a reduced fee. (In Britain, most AI centres are run by the Milk Marketing Board).

Milking

During the summer the cow or cows — two little ones will be better than one big one — will live outside day and night. Twice a day the milker will need to be milked and while this can be done in the field, it is usually more comfortable to tie her up in a cool, fly-free cowshed for this job. Flies cause restlessness and kicking during milking. Give the cow some concentrates at milking time unless it's against your principles, and she will look forward to her visits to the cowshed. She will relax while you get on with the milking.

Milking machines are not necessary for house cows unless you have something wrong with your fingers and can't milk by hand; there is so much work in washing up afterwards that it is quicker to milk one or two animals by hand.

Milking may seem hard at first, but once your muscles get used to it, the job is quite pleasant. Everything must be very clean; bucket, stool, hands. 'Hooded' pails are best if you can get them, they keep bits of hair and dirt from falling into the milk. Whatever bucket is used should be stainless steel and seamless so that it can be sterilised easily. It is particularly important to look after cleanliness in the dairy because milk is such a perfect breeding ground that any bacteria which fall into

it have a wonderful time; and it is not possible to make good dairy produce from bad milk.

Be firm and gentle at milking, and try to do the same things in the same sequence every time because cows like routine. Hold the teats in your fist and close your fingers one after the other from the top downwards, to force the milk down the tube. Milk is let down by the cow in response to stimulation, provided by the calf, or by an accustomed routine and washing the udder. The hormone oxytocin is released and this encourages the milk to flow. Once the process has been started, the job should be kept going — don't let anyone interrupt. Once the flow has ceased it is difficult to get it going again.

A great inhibiter of oxytocin is adrenalin, the hormone that makes your hair stand on end when you get a fright. Any sudden loud noise, even a shadow, noisy strangers coming in and so on can cause the cow to take fright and the milk flow will dry up. This means great patience is needed; if the cow kicks, don't kick it back. Try not to threaten it with a stick; be calm and quiet when you handle a cow and it should make all the difference, (this goes for goats as well). Few cows will kick for badness; most of them only do so when frightened, or in pain. If you get kicked look at the cow to see what's wrong. There may be a scratch on the teat that you have missed, or sometimes long hairs on the udder are pulled by the action of milking. Very hairy udders may have to be clipped in winter.

There are compulsive kickers and it is not always ill-treatment that makes them so. We have one at the moment, a healthy young cow with nothing wrong with her except a delight in kicking when being milked. If you can't win, put it in the freezer — cow beef is very good!

Lactation Curve. If the cow is getting the right food, her yield should gradually increase for about five weeks after calving. For five weeks after this it will stay about the same; then it is likely to fall again by 10% per month.

The quantity of milk she gives at the peak of her lactation, multiplied by 200, is approximately the total yield for the lactation. For example, a cow peaking at 3 gallons will probably give 600 gallons in total.

Some backyarders may prefer to produce milk from home grown foods and this may exclude most concentrates; but if you do feed them, it is usual to be generous with concentrates early on in the lactation to get the milk going well; at this stage your cow may be losing weight. This is quite usual in dairy breeds and is called 'milking off her back'. Dairy cows should not be too fat in any case, but keep an eye on her condition and don't let her get too thin. If she is using up body fat, this will provide energy but not protein, so she may be short of protein.

Feeding

Once again feeds are divided into maintenance and production. The maintenance ration is supposed to give enough food for the animal to perform all its functions with no loss or gain in weight. Fibre must be included, so that the rumen can work properly. A Jersey or other small breed of cow will need about 16½ lb of average hay or its equivalent as a maintenance ration. A general guide for dry matter appetite is to allow 2½ lb dry matter each day per 100 lb body weight. On free unrestricted summer grazing, she will get all this from grass. Winter feeds will vary — see the table of hay equivalents. If you want to feed bulky foods for production as well, the problem is the limit of the cows capacity; especially with a small breed. I estimate you could feed more hay equivalent to allow for one gallon or so of milk, but after this concentrates will be needed in winter. The hay equivalent should be 1:4 protein/carbohydrate ratio. Early summer grass is high in protein and will often provide enough nutrients for maintenance and production. Cattle cake is usually balanced to do this job and should be fed at 3 or 4 lb to the gallon according to the makers advice. Where the bulk food is thought to be low in protein, a high protein cake can be used to balance it.

HAY EQUIVALENTS

1 lb of average hay equals:	Protein/Carbohydrates Ratio
3 lbs medium silage	1:6
4 lbs Kale	1:5
4 lbs sugar beet tops	1:5
4½ lbs rape	1:4½
5 lbs mangolds (low protein)	1:15
1½ lbs oat or barley straw	1:20
2 lbs potatoes (no more than 20 lbs to be fed)	1:20
½ lb barley	
½ lb sugar beet pulp	
½ lb 'cow cake'	

That this is a rough and ready guide is illustrated by the fact that the proper ratio of protein to carbohydrate in a maintenance ration is 1:10, whereas the ratio in average hay is only 1:11. The last figure in the table gives the ratio of the other foods.

Minerals. Ask for advice on a new farm because the mineral deficiences are of the soil, which varies of course. Calcium and phosphorus are usually the critical ones with cows; salt licks are appreciated.

Calf Rearing

Modern dairy breeds of cow do not seem to have very well-developed maternal instincts; quite often this sort of cow will be indifferent to her calf and there will be no problem about weaning. The older breeds are more likely to be fiercely protective and could even be dangerous; so any newly calved cow with her calf should be approached with care, and never with a dog.

We will assume that the calf is left with the mother for the first day or two; this will enable it to get colostrum in what it considers to be the right quantities, and will also be good for the cow. For the first few days after calving, cows should not be milked right out to the last drop, but little and often, and this the calf will do. It is possible to induce a condition known as milk fever, which is a deficiency of calcium; the lack of it in

the blood causes the cow to sway on her feet, go down and go into a coma with her head characteristically turned backwards. The vet will inject calcium into her bloodstream and if he gets there in time, the effect is magical — instant recovery. It is interesting that this coma is similar to the hibernation sleep of some animals, as this winter condition is induced by a drop in the calcium level in the blood.

When the calf is taken away from the cow, put it out of sight and sound, in a smallish warm pen with deep bedding. Air temperatures for calves need not be too high, but there must be no draughts. The recommended temperature is about 55°F for the first month and about 45°F afterwards.

A calf bucket will be needed and the right amount of milk, warm from the mother or warmed up to blood heat with a little water, should be put into the bucket. The calf will suck your fingers and if they are lowered into the bucket it will soon be sucking milk; the beef breeds are a bit dim at first, but the dairy breeds of calf learn very quickly. Gradually over a day or two the calf can be taught to drink without your fingers, but at first it may get its nose into the milk and splutter a lot. The great difficulty is that the natural position of a drinking calf is with the head tipped back, while the milk runs straight down an oesophagal groove to the stomach; drinking with the head down is unnatural. A calf which has been left on the cow for a week or two is much more difficult to teach.

Calves are usually fed twice a day on the bucket, but if you have time they would appreciate a feed at midday as well. Regular feeding times are a good thing. A small calf such as a Jersey will take about 1½ pints of milk warmed up with ½ pint water twice a day, for the first week this will be all it wants. See the calf feeding table, which refers to Jerseys. Larger breeds will be bigger at birth and need more; a Friesian calf will drink about a gallon a day when it gets going.

Commercial farmers who sell milk usually change the calves over onto powdered milk substitute at about a week, unless they have any unsaleable milk. Backyarders will probably do better to keep the calf on its mothers milk; it will probably grow better and make a good use of what could well be surplus milk when a cow is giving three or four gallons a day.

A guide to feeding any calf is that it should get 12% of its own weight in milk every day. This works out quite well — if you can weigh your calf!

CALF REARING TABLE

Age	Milk	Other Needs
1–3 days	on mother	—
4–7 days	1½ pints milk + ½ pint hot water, mixed to blood heat twice a day	Hay, clean water, calf pellets all available. Put Elastrator ring on bulls.
1–3 weeks	2 pints milk + ½ pt water twice a day	Hay, water, pellets
3 weeks	Change to powdered milk if none spare	Hay, water, pellets
6 weeks	Larger calves will need 6 pints each feed	Dehorn when buds are felt change to rearing nuts
8 weeks	Can gradually wean off milk if eating plenty of solid food	Maximum concs 5 lbs per day, good hay ad lib
4–5 months		Can start feeding silage or roots

NB If calf scours, stop feeding milk. Give warm water and glucose for one feed, then ½ milk, ½ water for next. Comfrey is good — feed chopped in bottom of bucket.

10 sheep

Although they are naturally wide ranging, sheep of some breeds can be very useful in backyarding and fit into a balanced pattern of mixed stock. A glance at the table of sheep breeds will suggest that the mountain breeds should be avoided unless you have moor rights (access to hill grazing), because they are not happy in a confined space and they are diabolically ingenious when it comes to escaping. The lowland breeds have in the past been folded close together on arable crops and used in rotation round the farm, sometimes more for the fertility they left than for their own value. This could be a good idea for backyarders who wish to improve the fertility of poor land by the use of natural fertilisers. So small round lowland sheep, and the Jacob, an ornamental breed, are useful to backyarders, producing a good wool crop and small sweet joints of meat. Jacob wool is ideal for hand spinning.

There are many advantages in keeping a few sheep. They can graze after cattle and tidy up the pastures, nibbling down the grass shorter than the cattle because of their split lip. This and their manure means they are good for grassland management, provided they are not left on too long and allowed to overgraze. They must be given enough to eat; winter grass can be supplemented with hay and other foods such as kale and roots, and perhaps some concentrates. In former times, sheep were fed on rich food simply to improve the quality of the manure! Although this would not be worthwhile now, it does pay to keep the sheep well fed. If your grazing is very limited, perhaps it could be given a winter rest. It is traditional in some parts for sheep to be sent away to another farm for winter grazing, so a farmer might graze your ewes with his own for a winter, returning them in time for lambing.

Rewards from keeping sheep are wool, meat and milk. A fleece varies widely according to the breed, and also according to the sheep's health and circumstances; if a sheep is not healthy, the fleece will be poor. From most backyard sheep you can expect at least a 5 lb fleece every year, which will only be worth a pound or two if you sell it, but which can be spun and woven at home to make a useful garment or furnishing. Handspinning is a reviving craft and it should not be difficult to find out where to learn how to do it. Spinning wheels are expensive, but the more primitive tool, the spindle, is not. You can make your own spindle for a first attempt at spinning from a potato and a knitting needle; craft shops sell them fairly cheaply. When you are good at it, you can spin nearly as fast on a spindle as you can with the wheel — there is little difference, and the yarn is just as good. So lack of capital need not deter you from spinning your own wool, as it was done from five hundred years ago. Backyard Sheep Farming contains more details of the possibilites, Jacob wool is two coloured, but white wool can be dyed with vegetable dyes.

The meat from backyard sheep comes in the form of two lambs a year, or nearly so, from each ewe; unless some ewe lambs are kept for breeding and the flock increased this way, the lambs can make meat at any time from the late summer onwards. This can be frozen or salted down for the winter, perhaps 40 lb of meat from each animal, depending on the breed. In Britain, lambs are usually killed at 80 — 100 lb liveweight, to give a dressed carcase of 36 — 50 lb. Sometimes they are taken as 'stores' for further fattening and taken on to about

128

120 lb liveweight. This meat is economically produced; the lambs can be given concentrates to speed them up, but in the main they will produce meat from grass. The sheep is a ruminant and makes much better use of grass than the pig.

Milk from sheep is an old tradition; medieval backyarders relied on it for butter and cheese because the sheep was smaller than the cow and would live on poorer land. In Europe the tradition continued up to the present, but in Britain and America it had more or less vanished until about 10 years ago when some of the European breeds were introduced by people who were interested in self sufficiency. There are it seems few available at the moment, although I know of one source (see back of book, supplies section).

The East Friesland sheep comes from the same area of the world as the Friesian cow and seems to be likewise developed for milk production. This is a large breed with an impressive record; it usually produces two lambs (lambing percentage 200% in shepherds' jargon); the fleece weighs about 10 lb and the milk, at about 7% butterfat averages a yield of about 130 gallons, enough for a household. The milk makes rich dairy products and particularly rich cheese — Roquefort is made from sheeps milk. A lot of food is needed for all this production, and the sheep are rather delicate; one flock owner suggests that they would be better for the average backyarder if crossed with a hardier native breed.

Although in commercial sheep farming there is now a lot of specialisation, with lambs coming down from the hill farms to be fattened on the lowlands, backyarders at whatever point on the hill they live will want to see the whole process through, breeding, rearing and fattening. Unless, that is, you get too attached to your lambs and find you cannot eat them. This may happen if they are reared on the bottle, and many people find it best to get only female orphan lambs that can be kept for breeding. If the next generation prove equally difficult at least sheep are fetching a good price these days! They can be kept for wool or wool and milk, (you could in fact milk any breed if you accustomed it to the process) and the surplus sold for breeding.

Breeding

Sheep in general will only breed, ie take the ram, in the autumn, as the daylight length decreases and so affects their reproductive system. The ram is put with the ewes then, and the lambs will be born 21 weeks after the ewe is served. It is better if the lambs are not born too early in the Spring; warmer weather makes a great deal of difference and we can plan for April lambs, whereas commercial lamb producers have to start earlier to catch the market. Ewes are in season for up to 36 hours at 14 — 21 day intervals.

It may be possible to borrow a neighbours' ram, or put your ewes in the field with his ewes for a few weeks. It will not be worth keeping a ram of your own for a very small backyard flock; he will not be economic. Fortunately, sheep farmers have a long tradition of helping each other, and in most districts there should be no problem about finding a ram. Try to borrow a good one, whatever breed it is. The most usual breed in lowland flocks is a Down breed, bred for fat lamb production.

Just a week or two before they go to the ram, the ewes are 'flushed' which means putting them onto some better grass so that they are improving a little in condition, (after being fairly lean) when they meet the ram. Very fat ewes will not breed so well. Very woolly ewes should be clipped a little round the tail for a more efficient service.

During the first two thirds of the pregnancy, the ewes will probably need little extra food to what they can pick up, although where their grazing is very restricted they may need it; but a lean condition is best until late in pregnancy when they should be given more (see feeding).

130

Lambing. Most shepherds try to be there when their ewes are lambing, in case anything goes wrong. A shed or temporary shelter made of straw bales will be a great help. When the ewe lambs, clip the wool away from the teats if necessary and make sure the milk is flowing, and then stand by to make sure that the lamb can and does suck and that the ewe looks after the lamb properly. Care at this time can save lives! Sometimes a ewe will neglect her lamb and if it gets chilled it will soon die. A ewe with too much milk should be milked out by hand to prevent the onset of mastitis, a painful inflammation of the udder.

Experienced shepherds often assist their ewes with lambing; a lamb can get twisted on the way out and it needs a gentle but sure manipulation to help it into the world. Your vet or a shepherd friend should be called in if a ewe seems to be in difficulties, and if you watch what happens you will build up your own stock of veterinary knowledge.

131

SOME BREEDS OF SHEEP

Breed	Advantages	Disadvantages
Jacob spotted, with 2,4 or 6 horns. Were rare, becoming more widespread for backyarding	Docile, good for small areas Hardy, fairly prolific Wool brown and white and good for hand spinning Meat good – small joints with little fat Very economical producers of meat Long-lived	May be more expensive than commercial sheep Horns can be a problem
Hill Breeds Scottish Blackface Welsh Mountain Swaledale, Rough Fell Cheviot etc.	Bred for hill conditions, so if you have upland grazing, use the local breed or cross of Hill sheep. Hardy and adapted to their homelands. Small joints of excellent meat from land that would not support cattle	Light fleeces with large amounts of kemp Difficult to shepherd unless you are an expert *Not* good in confined spaces. Born escapers Horns can be a problem
Longwool Breeds Border Leicester Leicester, Wensleydale Lincoln Longwool Cotswold, Romney Marsh Dartmoor etc	Very heavy fleece with long staple Wool lustrous – has sheen Large carcase Adapted for lowlands, docile No horns	Coarse wool Too big for confined area Meat rather fat and joints large
Down Breeds Dorset Horn, Southdown Shropshire, Dorset Down Hampshire Down, Suffolk etc.	Very early maturing lambs for quick meat production. Good quality meat and wool. Mostly hornless Fairly docile and will adapt to limited area	Not hardy enough for adverse conditions Fleeces rather light in weight
Grassland Breeds Kerry Hill, Clun Forest Radnor	These look like Down sheep, but they graze moorland on the Welsh border Good for moors	More flighty than Down breeds Local to Welsh Border as rule
Milk Sheep e.g. East Friesland	Yield of milk nearly 300 gallons 7% butter fat Prolific, two lambs per ewe Good fleece. Docile, used to handling	May be difficult to find Eats a lot!

Rearing. The easiest way to get ewes of your own tame and
docile is to get hold of female orphan lambs and rear them on
the bottle, which is tedious, but worthwile. They are usually
available at lambing time in sheep districts because there is often
a surplus of motherless lambs in commercial flocks. If you have
a goat or cow there will be milk at hand, or if not there are
special milk powder mixtures formulated for lambs — Ewelac
is the trade name of one.

Try to get lambs which have had colostrum from their mothers — the first milk, containing extra vitamins and antibodies.
Without this they will be difficult or impossible to rear. There
are recipes for a colostrum replacer, sort of egg nogs with a dash
of brandy, but nothing can give antibody protection except
the mothers' milk. The vet may be able to provide a specially
medicated first feed to help to get over this problem.

It is probably better to bottle feed lambs than to allow them
to suck the goat, although I know people who have put them
straight onto the goat, holding them and waiting for the
lambs' little stomachs to change from hollow to straight-sided
before taking them off again. But if you milk the goat first,
and then feed the lambs from a bottle, you can regulate exactly
the amount they take and avoid possible damage to the goat's
teats. Lambs are rougher with the teat than kids, so do watch
for damage if you allow a goat to suckle lambs.

When you take home a new lamb, keep it warm; a lamb box
lined with hay should be kept over a radiator or near an all-
nightstove, or a hot water bottle could be put in with the lamb.

You may have to teach it to drink from a bottle. Try to get
the proper lamb's teat for the purpose as it makes the job a lot
easier. Keep the milk at blood heat and follow the instructions
if you use a milk powder. For a weakly lamb, a little honey and
cod liver oil can be added.

The rearers who are most successful say they start with very
frequent feeding — say 2 oz of milk every 2 hours. After a few
days this can be spaced out to four hour intervals. A vigorous
lamb will soon get through the night with a late night and an
early morning feed. If the lamb scours, cut down the food and
in a bad case, feed warm water and glucose instead of milk.
Arrowroot or slippery elm food are good for scour; they are
mixed with watered milk.

Let the lambs out on to grass in good weather after about a
fortnight, and by then they should have available water and

lamb pellets. It will help the lambs if they can get onto solid food as soon as possible, and grass will encourage the development of the rumen. This development has to take place gradually before weaning. They can be weaned when they are visibly eating plenty of solid food.

There are one or two other points to watch with new lambs. Navels that have not dried should be dabbed with iodine. Male lambs should be castrated with the rubber ring called an 'elastrator' and it is illegal to use this after the lamb is one week old. The tails of lowland sheep are also docked very early with the same rubber ring. It is not a pleasant job, but in lowland districts the tails are taken off to protect the lamb in later life from the horrible blowfly, which may lay eggs under the tail. Upland sheep usually keep their tails since there is less danger from fly in the hills.

General Management

See your sheep at least once every day and really look at them to see that all is well; prompt action can anticipate trouble. Sometimes they get 'rigged' — fall onto their backs, lie helpless and will die if not righted. Very often your approach will give them the extra impetus they need to get up.

Blowfly is a sheep scourge in summer. It is a green fly like a small bluebottle which lays its eggs on the sheep; if the maggots hatch out they literally eat the sheep alive. A sheep which is not in the best of health seems to be more likely to be 'struck'. A watch is therefore kept for the eggs, and they are removed by clipping the wool; and as a precaution, the wool round the tail is clipped fairly frequently in warm weather. This is called 'dagging'. If any maggots hatch, they must be picked out and the wound treated with antiseptic.

Make sure that all your sheep have water to drink. You will find that they drink quite a lot, especially in dry weather. They will need shade in summer and shelter in winter, and if there is no natural shelter, perhaps a straw baled hut could be rigged up or they could have access to a shed.

Footrot. Part of general management is watching the feet of your sheep. By nature the sheep is equipped to travel long distances on stony ground, and this will keep the feet in trim; but on soft lowland pastures the feet get soft and are sometimes invaded by the footrot germ and other bacteria, which can cause lameness.

It starts with a slight inflammation between the toes; later the horn separates from the hoof. Pastures are contaminated by the germs, but they will be clean again provided that the sheep can be kept off for fourteen days — another good reason for the use of paddock grazing. Before going onto clean pasture, and as a routine where the disease is present, sheep should be walked through a foot bath containing 5% copper sulphate or formalin.

Affected animals should be treated at once; with a sharp knife the diseased tissue is cut away and the hoof treated with one of a variety of things. Some farmers swear by Stockholm tar for this job, others dip the foot for ten seconds in a 30% copper sulphate solution. Another treatment is a mixture of butter of antimony and glycerine, applied to the foot at three day intervals with a quill.

If you catch a sheep with suspected footrot and feel the feet, an affected foot will feel hotter than the others. Paring the feet is quite a skilled job, but you can soon learn by watching an expert.

Shepherds say that short grass is a good insurance against footrot; where sheep graze long grass it is sometimes dragged through the cleft in the foot, making it sore and allowing the bacteria to enter.

Feeding

Grass will normally be all that your sheep will need, once the lambs are reared, up to Christmas — although to fatten the remaining lambs more quickly some concentrates can be fed. Backyard sheep thrive on a varied diet of garden waste, so if grass is short they will be happy with other greenfood. After Christmas when there is little grass, and the ewes are in lamb, they will need good hay in racks and some concentrates. There are sheep nuts and sheep meal mixtures sold by feed merchants, or you can make up your own — possibly from home grown cereals. The sort of ration sheep farmers use might be 4 parts crushed oats, 1 part flaked maize and 1 part linseed or groundnut cake. For a higher protein ration, reduce the amount of oats. Feed ewes from about ¾ lb of this per day at first, to an absolute maximum of 1½ lb per ewe in a clean trough. Feed according to common sense; the ewes must not get too fat or they may have a difficult lambing, but they must be properly fed or they might get twin lamb disease. In a hard winter they will need more food to keep warm, as will any type of stock.

Where there are other foods instead of grass, the sheep have to be rationed sensibly. A ewe can eat up to 12 lb of silage a day; it is a good food for sheep. The general rule is that about 3 lb of silage equals 1 lb of hay. Turnips and swedes were traditionally grown for sheep; they can be eaten off in the field if it is well fenced, or carted to the sheep. They are fairly frost hardy.

Kale may be grown for several types of stock and sheep will enjoy it; a suckling ewe can tuck away as much as 20 lb of kale a day, or the same quantity of swedes. The ration could be completed with about 1 lb of hay and perhaps 1 lb of concentrates as well. Fattening sheep will eat about 10 lb of kale a day. Fodder beet and mangolds are other useful sheep feeds, and fodder beet or sugar beet tops, after wilting, make a good bite.

Shearing

This usually takes place in May or June, but the time varies with place and season. Sometimes the date of shearing is hurried forward because the flies are about which cause blowfly strike, and shorn sheep are less likely to be attacked. The warmer weather causes the grease to rise in the wool, and the wool itself rises slightly off the skin, and this makes shearing easier. It is a skilled job and a hard one, whether done by hand shears or electrical clippers. It would be very satisfying to learn to shear your own sheep, but watch an experienced shearer first and shear several sheep under his eye before you do it on your own. Go too close to the skin and you will cut the sheep; too far away and a double cut will be necessary. If the sheep is sitting comfortably in the proper position, she should not struggle.

It is not just for their own convenience that shepherds choose a fine day for shearing. If the wool is shorn when it is wet with rain, dew or sweat, it will not store well; it will be musty and perhaps discoloured when the fleece is unrolled. On a very hot day, the sheep waiting to be sheared should be allowed some shade and they should not be hurried before shearing. It is of course better for the newly-shorn sheep if the weather is mild.

In Britain, owners of more than four fleeces who wish to sell them should register with the Wool Marketing Board.

Dipping

Sheep should be dipped in summer whether it is compulsory or

not. At the time of writing, sheep scab is present in Britain and there are regulations about dipping sheep to control this disease, which is an extremely unpleasant skin irritation caused by a mite. The dip must be a product likely to kill mites and be approved by the Ministry of Agriculture for the purpose. For those reluctant to use the conventional sheep dips, which contain BHC (banned in some countries) there is a list of alternative dips in Backyard Sheep Farming. These have been approved long ago, and have been brought back after a great deal of trouble and protest by some farmers. They may not have been heard of as acceptable alternatives in your part of the world, but the County Council Special Officer will have to be informed when the dipping is to take place so that he can be there if he wants to; and so you could at the same time discuss the type of dip you intend to use and get permission for it. The more 'organic' dips have to be used twice, at eight-day intervals.

Apart from scab, shepherds dip their sheep to discourage other external parasites, not so deadly as scab but very uncomfortable for the sheep. Keds and lice are two of these; the sheep can be dipped as soon as the wool has grown long enough to hold the dip, say two or three weeks after shearing, and then again later in the summer. Fly strike is also discouraged by dipping, so the sheep could be dipped more frequently in a bad blow-fly year. The effort of arranging a dip is well worth it for the sheeps' comfort and wellbeing. Many backyarders do the job on the lawn, in a tin bath, suitably attired in oilskins and with a willing helper on hand. Others join a farmer when he dips his own flock, in a proper dip with all the sheep handling facilities.

Hot, panting sheep should never be dipped; so if they have to walk to the dip, give them a rest before they take the plunge. Wet weather dilutes the dip rather too quickly, so try to choose a fine day.

THE VOCABULARY OF SHEPHERDS

Sheep names vary according to district. Even the word 'lamb' is rather vague, although strictly it belongs to an ovine animal that has not got its first pair of permanent incisors.

Females

Ewe lambs, gimmer or chilver	— birth to weaning
Ewe teg, gimmer hog, ewe hog	— weaning to first shearing
Shearing ewe, shearing gimmer gimmer or theave	— first to second shearing
Ewe	— has had a least one lamb
Crone	— an old barren ewe

Males

Ram lambs, hoggets, hogs	— weaning to first shearing
Shearing tups or rams	— after first shearing
2, 3, 4 shear rams or tups	— according to number of times they have been shorn

Castrated Males

Wether or wedder	— weaning to first shearing
Shearing wether	— first to second shearing

11 animal health

Sir Albert Howard, the compost pioneer, said that diseases of plants and animals are Nature's professors of agriculture, to keep us up to the mark — indications that something is wrong in our management. This is about as far as you can get from the germ theory of disease, but it seems to be a constructive idea. It is the basis of the organic school of thought.

When Pasteur discovered that certain diseases occur when certain germs are present, this seemed to explain everything, but now it seems that things were rather over-simplified. It was thought that if you kept away the wicked pathogens, everything would be healthy. But there have to be predisposing conditions before the bugs can get a hold, and these are now thought by many authorities to be the key. The trouble is that these conditions are not easy to identify.

The theory fits in well with the fact that bacteria and viruses are often present harmlessly in healthy animals — for example the coccidiosis germ. Then something happens to upset the balance and the germs assume control, causing illness in the host. This is not always in their own interests, because if the host dies they die out. Just what goes wrong is not known, but it seems that the difference can be described as stress. When the animal or plant suffers stress, its natural resistance is lowered.

In practice this means that we can have a positive approach to our own health and that of our crops and stock. We need to take active steps to prevent disease, rather than curing specific ailments with specific remedies.

In the tables of this chapter I have listed a few of the more common ailments of the various types of stock, and the usual remedies, but the first question should be — what went wrong? What was the predisposing factor? Treat the ailments, but try to find the underlying reason for them so as to eliminate it.

Our first positive step involves a healthy respect for the bacterial causes of disease. It is best to try to keep animals and their surroundings reasonably clean, for their comfort as well as health. Sensible precautions include giving a clean pen to a sow about to farrow or a goat about to kid; unless they are able to give birth outdoors, which will probably be better in mild weather (on our farm all the calves are born outside in summer). On the plant side, there are also precautions — diseased plants should be taken away and burned, not passed round from hand to hand for diagnosis.

Blanket antibiotic treatment I would never include in the list of sensible precautions. Poultry, rabbit and pig foods are often laced with various drugs in order to keep down disease in large scale enterprises, where they are probably necessary.

Pig growercare pellets

At the time of manufacture, Vitamins A, D$_3$ and E were added to this feed. The following levels are guaranteed until the end of the stated month.

Vitamin A 10,000 i.u/kg   Vitamin D$_3$ 2,000 i.u/kg   Vitamin E 10.0 i.u/kg

Contains a permitted anti-oxidant.

FOR ANIMAL FEEDING ONLY   For feeding to young growing pigs

DO NOT FEED TO ADULT BREEDING STOCK, POULTRY, RUMINANTS, OR PIGS OVER 26 WEEKS OF AGE

700 mg/kg COPPER SULPHATE (PL 2987/4000) was incorporated as a mineral nutrient. Total Copper 200 mg/kg.

300 mg/kg AVATON 50 (15 mg/kg Avoparcin PL 0095/4026) was incorporated to improve growth rate and feed conversion efficiency and

200 mg/kg QUIXALUD Premix (120 mg/kg HALQUINOL PL 0034/4001) was added to minimise the occurrence of scours caused or complicated by E. Coli and Salmonella spp., and non-specific scours during the growing period, eg. 6-12 weeks of age The above were added at the time of manufacture and during the period when they were expected to retain their potency.

Care should be taken to see that sheep do not have access to effluent from pigs fed copper supplemented diets,

It should never be needed in backyarding, because there is not so much financial risk where there are small numbers of animals and also because the more natural methods of feeding and management will probably reduce the risk of serious infection.

Big outfits carry their own forms of stress. One example of this is the common occurrence of the loss of large numbers of poultry during a power cut, when the forced ventilation system breaks down. Backyard poultry keepers never encounter this sort of problem, thank goodness.

When it comes to disease, the organic approach is to build up a natural immunity so as to give plants and animals resistance of their own; and this rules out blanket treatments, because these will destroy natural resistance. This is quite apparent in the world of crops where those that have been sprayed, say against aphids — suffer from a second wave of the pest, much worse than the unsprayed crops, which have gained immunity during the first attack.

When the biological cycle is allowed to operate intact, the result is a healthy soil and the plants and animals which depend on that soil are likely to be healthy too. They may not be the largest or highest yielding specimens, but what they lack in quantity will be made up for in quality. This is an aspect which has been badly neglected by commercial interests, which tend to judge everything by eye for its sales appeal.

One example of organic thinking is mixed stocking, where pastures are grazed by a variety of animals. This is beneficial to health because worms are usually specific to one species and it interrupts the life cycle of the worms when another species gets hold of them. Internal parasites are a problem on heavily grazed land and this is one way round it. On a garden scale, this can be achieved by grazing a patch with rabbits and then changing to poultry for the next grazing.

Less obvious but becoming clearer with the years is the fact that plants and animals of varying species differ in their requirements. They will take from and return to the land different elements which should avoid the danger of robbing the land of too much of any one thing.

Organic methods minimise stress, because plants and animals are allowed to grow naturally and are never pushed to the limit. In day to day management stress is also to be avoided; and this means being on the watch all the time. Observe animals and crops! Visit them regularly — an example of stress to all living things is going without water. Build up a picture of your animals' behaviour, because only when you really know how they normally behave can you develop that sixth sense which tells you when something is wrong. An early diagnosis of trouble is worth a great deal; an animal can often be saved by prompt action.

It is worth saying again that all changes should be made gradually, because natural things hate sudden change. If your cow has been spending most of the winter indoors, let her out for short periods at first until she gets used to the change of temperature and food. Let all animals have very restricted access to fresh spring grass, for short periods at first, because lush grass after winter feed can cause scouring and even bloat. Keep feeding a little hay for the first week or two.

Newly arrived animals will be suffering from shock because of the journey and the change of surroundings. One way to minimise this is to find out what has been fed and to feed the same ration for a while. Slowly you can make the change to your own ideas of feeding. Mix in new animals with your own only after they have had some time to settle down and get over the change.

New animals particularly, and all animals in general, should be watched for abnormal droppings, usually the first sign that all is not well. This is more obvious when animals are indoors than when they are out in the field, but it is an important guide to health.

Regular hours are quite important to animals. It seems to me that one of the reasons the monks were good farmers may have been their orderly lives. Summoned by bells, they and their helpers were working to the clock and the animals were bound to get regular feeds and attention. This may seem irksome with a backyard enterprise, where you are actually trying to get away from the regimentation of modern life and please yourself, but you will find that animals impose their own discipline. It is very important to milk cows and goats regularly, particularly when they are yielding heavily; it is downright cruel to leave such an animal with a tightly packed udder, waiting to be milked until you come back from a night out. Twelve hour intervals between milkings are ideal, but they are too rigorous for most of us. The animals will get used to your arrangements, whatever they are, so long as they are regular. Irregular milking hours constitute stress.

PARASITES

These can be internal and external, living things on the animals' body which carry out their life cycle, or part of it, on their hosts. Some of the life cycles are very complicated indeed. When present in large numbers, parasites can cause discomfort, illness and even death. They suck blood, block up organs and eat food meant for the host; but it is unrealistic to expect to wipe them out completely. We try to keep them down to a reasonable level and to maintain the health of the animals so that they have a natural resistance. It is well known that a sheep for example that is underfed or not well is more likely to suffer from an attack of worms; the organism has got out of balance.

Internal Parasites

Stomach Worms. Cattle, goats, sheep and pigs are all affected by these but they are specific — cattle worms do not affect sheep, and so on. They vary in number according to the time of year and the intensity of stocking on the pastures, and the weather. Eggs fall to the ground with the dung, and if the weather is warm and damp, many of them hatch out in a few days and become larvae. At this stage they may be picked up by a grazing animal, when they will stay in the stomach of that animal and grow into adults. (Eating them in the egg stage does no harm; they are digested.) This is why pastures become progressively dangerous if they are grazed too often by the same species of animal.

Three or four weeks between grazings will normally be enough to allow the larvae to die because they do not survive longer than this without finding an animal host. The great advantage of grazing animals in paddocks with frequent moves is that most of the ordinary round worms will have died out by the time the patch is grazed again.

Lung worms. These can cause Husk or parasitic pneumonia; another species cases bronchitis. The worms live in the air tubes of the lungs. The larvae are coughed up, swallowed and go out to the pasture in the dung. After about a week they will infect other animals which pick them up with the grass. They are swallowed, get into the intestine lining and are taken to the lungs by the bloodstream. If not picked up, they can live on the pasture for about a year.

Pigs do not get lung worms unless they are out on pasture; the most commonly affected stock seem to be young cattle. The symptoms of fast breathing, followed by coughing, are usually noticed in late summer and early autumn. If they cough a lot when you make them run around they probably have lung worms. The heavy dews of autumn give the larvae their favourite conditions, so it helps if calves are brought in at night from late August onwards. A few of these worms give young cattle immunity; in fact the treatment many people give to their calves is a live vaccine called Dictol. It consists of a dose of larvae, irradiated by X-rays. They cannot damage the animal, but they cause it to produce antitoxins which will immunise it from further attack.

Liver Fluke. This parasite is a nuisance in wet seasons. It is particularly dangerous to sheep, but cattle, goats and rabbits can be affected. Grazing animals on dry and well drained

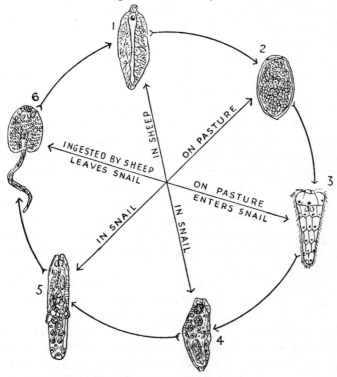

LIFE CYCLE OF LIVER FLUKE

1. Adult Fluke.
2. Egg of Fluke.
3. Miracidium.
4. Sporocyst.
5. Redia.
6. Cercaria.

ground will avoid this problem, but this is not always possible. The fluke passes at one stage of its life cycle through its alternate host, the water snail *Limnae truncatula.* In bad cases farmers sometimes try to kill off the snails with chemicals such as copper suphate, but chemicals may be dangerous to the stock. Ducks are a better form of control, they will gobble up the snails; these creatures live in areas which are halfway between dry ground and swamp.

Animals affected by liver fluke develop anaemia and look pale and miserable; goats sometime show a swelling under the jaw. Cows on wet pastures are sometimes affected.

The herbal treatment for liver fluke is as for worms, garlic particularly. Chopped dandelion in the diet helps recovery. In a bad attack, it may be necessary to dose with one of the modern injections such as Zanil, made by ICI, or Trodax, by May and Baker.

We have been warned several times about people being able to pick up liver fluke from eating watercress upon which the cysts are lurking.

In 'flukey' areas it may be wise to fence off or even drain wet patches of land which harbour the snail; but you will find that seasons vary enormously.

Warble Fly. This is a parasite of cattle which looks like a bee, and hovers over its victims in a menacing way; cattle become alarmed and they 'gad,' ie run about with their tails in the air in a sort of panic. The fly lays its eggs on the hairs on the undersides and the legs of cattle. The larvae hatch out and eat their way right through the animal, by the next spring arriving at the animal's back, and there they bore a breathing hole through the skin. They grow on the backs of cattle to form big lumps and then when ready, drop out and fall to the ground, where they form pupae. The adult flies hatch out from these and the cycle goes on.

The leather of the skin is ruined by warble flies, and they are quite common. The animal must suffer quite a lot and it is clearly our duty to do something about it if we can. It used to be compulsory in Britain to dress the backs of cattle with a solution made of derris powder. This is rubbed into the back and if it gets into the holes it will kill the maggots before they emerge — not in time to help this season, but reducing the warble fly population. Because of this long term effect, farmers dropped the treatment when it was not legally enforced and not many people use derris now. The modern treatment is to apply organo-phosphorus dressings or drenches (doses by mouth), but these are dangerous. They get into the blood-stream and kill the larvae at that stage; but an excess dose can be fatal to the cattle as well. It is usually given in autumn, whereas derris is used every month from March to June. The scabs must be removed from the holes for the derris dressing to be really effective, so it has to be really scrubbed in and not merely swabbed over. Powdered derris and soap is the best mixture; but derris is dangerous to fish, so keep the warble dressing away from your fish pond.

External Parasites

Lice. These again are specific to the various types of animal, for example, there are four sorts of lice that can infest cattle and three for goats. Some are biting lice and some sucking. They cause irritation to the hosts, which leads to scratching, loss of hair and eventually loss of condition. They can be seen by the naked eye, so look out for them on restless animals, particularly those indoors at the end of winter. Modern louse powders are most effective, but they contain BHC. Old fash-ioned remedies include waste oil from garages, which works well on pigs, derris powder, and tobacco. Hens are sometimes prone to lice and other parasites such as red mite, and I have seen a useful gadget to prevent this. It is a hollow egg made of pot, which was filled with tobacco and then put in the nestbox where the warmth of the hen brought out fumes to kill the parasites in the feathers. Mrs. Levy the herbalist sells an insect powder made from organic ingredients which gets rid of ex-ternal parasites.

Mange. Mange mites are smaller than lice and cannot be seen, but their effect is just as unpleasant. Pigs, for example, go red and lose their hair, and gradually the skin of affected animals becomes crusty and thickened. Hoescht make a dress-ing called Alugan for this which is very effective. For mild cases, try the herbal insecticide or a rubbing over with lemon peel or lemon juice.

Ticks and keds are unpleasant parasites which plague sheep. They can be controlled by a series of dippings during the summer months, and when sheep are handled they should be examined for these creatures. Another problem with ticks, this time with cattle, is that in some districts they carry a disease called red water which they transmit to the cattle. If

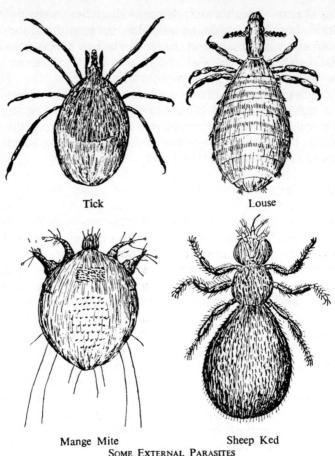

Tick	Louse
Mange Mite	Sheep Ked

SOME EXTERNAL PARASITES

one of your cattle starts to produce urine with a reddish tinge, call the vet.

Ringworm. This is an unpleasant disease caused by a fungal growth on the skin. It can occur in any animal including of course man, but is probably most usually seen in young cattle that have wintered indoors, in old buildings. It seems that the spores lodge in crevices in old woodwork. The animals develop their own immunity in time and mild cases will right themselves when they are put out to grass, but the risk of your being infected is such that this condition should be treated. In any case, a badly infected animal is a sorry sight and things may spread.

Treatment for this as for so many things, is usually an

aerosol spray from the vet; there is an additive to the feed which also helps. The standard treatment before the ubiquitous aerosol was to wash down with soapy water containing 2% formalin. Use rubber gloves for this job. Don't scrub them too much as this may spread the infection.

Another recommended treatment for ringworm is washing with a 10% solution of soda, followed by painting the lesions with iodine. Herbalists use house leek and also Herb Robert, used as a wash. There is an antibiotic treatment for ringworm now and it would probably be justified in severe cases.

The best book for those wishing to follow herbal treatments is J de B Levy's Herbal Handbook for Farm and Stable, which contains almost everything you need to know. I have listed a few of the diseases likely to occur in the tables, but the subject has not been fully covered by this chapter. You will build up your own stock of knowledge from experience, listening to the vet's advice and calling him before things get too serious. Money spent on vet advice is never wasted; for simple things in the beginning, after which you can rely on your own judgment and use the vet for serious cases. Never be afraid to ask questions, but do remember to listen to the answers.

Notifiable Diseases

These are diseases which are potentially so dangerous that responsible people such as owners of animals and their vets are obliged to report an outbreak to the police. They and the MAFF Health Department then arrange for animal movement restrictions in the area to try to prevent an outbreak from spreading. They include :
- — Anthrax in cattle, sheep, horses and pigs
- — Foot and mouth in cattle, sheep, goats and pigs
- — Swine fever in pigs
- — Bovine TB in cattle
- — Sheep scab
- — Fowl pest
- — Teschen disease in pigs

SOME DISEASES OF POULTRY

Disease	Symptoms and Causes	Remedies
Coccidiosis In chicks	Blood in droppings; chicks droop with eyes closed	Isolate in warm cage with no food, only water for 24 hours. Then feed laxative diet and 1 clove garlic daily for 10 days
BWD (Bacillary White Diarrhoea)	Chicks look chilled, white diarrhoea, many die	Cull and disinfect premises with formalin
Feather picking, cannibalism	Boredom, or parasites or lack of protein	Isolate and dress wounds with Stockholm tar
Catarrh	Cold — may be draughts: check chicken home	Dose with cod liver oil. Give plenty of fresh greenfood. Move to fresh ground
Worms in droppings	Loss of condition, scour	Laxative diet, plus garlic or tobacco in water
Egg bound Layers' Cramp	Hen tries to lay but is unable Squats with one leg cramped	Dose with castor oil. Massage vent with olive oil and heat over hot water
Sourcrop	Distended soft crop, movements of neck	1 teaspoon olive oil, then massage crop. Fast 24 hours
Gapeworm in windpipe	Coughing, sneezing, shaking head, gasping	Move to new ground. Isolate, fast 24 hours, then laxative diet plus garlic for 10 days
Lice, mites, fleas	Irritation, loss of condition	Isolate, give dust bath with Derris powder added Clean out premises and whitewash

In intensive poultry units, the only treatment for sick birds is to cull them. Backyarders usually have a try at curing them first.

SOME DISEASES OF RABBITS

Disease	Symptoms and Causes	Remedies
Scour commonest reason for loss	Diarrhoea and loss of condition caused by overfeeding, wet food, frosted or bad food	Cut out food except for hay and water plus a few herbs: blackberry leaves, shepherds purse, burnet, plantain. Check feed routine
Sneezes, Colds, Chills	Respiratory troubles, rhinitis or pneumonia May be infectious but often caused by draughts or bad ventilation	Check ventilation. Usually worst in indoor rabbits, get outside if possible and isolate from other animals. Cull if no improvement. Avoid direct draughts
Enteritis	Internal irritation, often due to high pressure commercial feeding. Often in young rabbits Lose appetite, are thirsty, grind their teeth and pass clear dung	Natural diet will help. Bread and milk plus hay only at first
Coccidiosis	Two kinds, affecting intestines and liver. Rabbits stop growing or with second kind, die	Garlic is a great cure for this. Vet will prescribe antibiotics and if there is an outbreak, may be justified
Ear Canker	Inflammation: rabbit scratches ear and shakes head. Caused by mites, which can live away from rabbit for several weeks, so disinfect hutches	Neat witch hazel applied with cotton wool to soften deposit. Next day try to remove it gently. OR benzyl benzoate at 5 day intervals
Sores	Caused by fighting, or on feet may be wet or rough cage floors	White clover. Separate fighters. Give clean, soft bedding. Use soothing ointment

SOME DISEASES OF PIGS

Disease	Symptoms	Remedies
Erysipelas	Can affect pigs of all ages. Lose appetite, depressed, constipated. May die if not treated. Happens more when weather warm and humid. May be lame, with swollen joints. May have diamond-shaped red patches on skin. Can affect heart.	Crystal violet vaccine is good prevention and it may have to be regular if disease appears. Get the vet
Necrotic Enteritis caused by coliform organisms	Young pigs in cold and damp quarters are most susceptible. Bowels infected with bacteria and wall of intestine destroyed. Diarrhoea appears — scour is greyish in appearance. Pigs are thin, with long hair and arched backs	Clean dry housing and warmth. Gradual changes in diet. Antibiotics in extreme cases
Virus Pneumonia	More prone to this when they have a lot of worms. Pigs sneeze and cough a lot. A disease of indoor pigs	Can clean empty pens with washing soda and virus will be dead in 2 days. Good food and ventilation — or turn them out, with a warm dry bed for sleeping
Oedema	Affects young pigs 5–16 weeks often biggest in litter. Change of some kind triggers it off. Pigs die suddenly, some stagger, have swollen eyes	Changes gradually — no sudden changes. New pigs given little food at first. If oedema suspected starve pigs for 24 hours, give dose of Epsom salts
Piglet Anaemia	Small sucking piglets reared indoors can be affected. Scour and pigs pale and thin	Treat all indoor piglets with iron at 2 days old. Sows and gilts outside will pass on some iron to piglets. Throw in turf to indoor sows

SOME CATTLE AILMENTS

Trouble	Symptoms and Causes	Remedies
Mastitis inflammation of the udder	Stress, fighting etc. causes the bacteria to attack. Clots in milk, inflammation of udder	Get rid of chronic cases. Remove stress. Check that milking is efficient. Treat with intramammary antibiotic injections. Herbal treatment, let calf suck cow, Wood Sage is recommended — see Mrs Levy for details of brew
Milk Fever (hypocalcaemia)	Usually affects newly calved cow. She 'paddles', sways and goes down, often with head turned to one side. Coma follows, and death if not treated. Due to blood calcium too low. More likely in older cows, sleeping out in autumn months	Get blood calcium up again and recovery is swift. Calcium borogluconate can be injected subcutaneously OR the vet will do it intravenously. Feed molasses, watercress
Grass Staggers (hypomagnesaemia)	Fall in level of blood magnesium, often in spring when Mg content of grass is low Cow may die suddenly. Sometimes twitchy and nervous. A disease of large herds	Can treat with magnesium salts in solution, injected with flutter valve. Feed extra Mg in rations can buy cow concentrate with Mg supplement Less likely on herbal leys
Acetonaemia	Milk goes down, cow loses appetite Breath smells of pear-drops. Usually in winter on dry rations. Happens more often when cow too fat at calving	Vets give glucose Try to find some green food — turning out to grass is best cure
Bloat	Worst on spring grass, especially clover Can happen if they break through electric fence. Animal swells up with gas and can die if not relieved. Caused by froth in rumen which prevents gas escaping in belch. ALWAYS make change to greenfood gradual; feed hay in morning before going out to grass in spring	Push stiff polythene tube straight up posterior, gag of hay in mouth to promote chewing. Dose with oil or even milk to lower surface tension. In emergency, puncture rumen with sharp knife. Get vet to show you exact spot, just in case!
Brucellosis (contagious abortion)	Brucella abortus is germ present. Abortion occurs 5th to 8th month. Foetus can spread infection. Can blood test for it	Cull infected cattle, as this germ causes undulant fever in man

153

SOME DISEASES OF SHEEP

Disease	Symptoms	Remedies
Worms lungworms and stomach worms. Picked up from pasture. Lambs particularly vulnerable	Particularly Spring and early Summer Unthrifty appearance, watery eyes, coughing and scouring	Fresh ground plus treatment. Garlic and molasses — chop up garlic leaves and mix with bran and molasses. In emergencies, a linseed and turpentine drench can be given. Adult dose is 80 drops turpentine in 2 ozs linseed oil. Lamb dose is 2 oz mustard seed
Scour	Droppings wet. Wool soiled caused by over-rich grazing or worms	Laxative to clear out intestines. Treat worms as above, or if nutritional, cut down grazing and feed hay. Slippery elm bark dose
Blowfly Strike	Eggs in clusters in wool May—September. Sheep uneasy when maggots hatch	Clip soiled wool around tail. Dip in early summer Keep strict watch on all sheep and remove any eggs
Sheep Scab	Irritation causes scratching, wool drops out. Sheep are restless, rub and bite fleece	Dipping frequently with an insecticide dip. Most approved dips contain BHC but there are less toxic ones which may be used
Footrot	Animals lame, grazing on knees. Bacteria is Fusiformis nodosis which survives about a fortnight in damp pastures	Pare down hoof and cut out dead tissue. Treat with 10% formalin or Stockholm tar. Run flock through copper sulphate or formalin footbath every 6 weeks. Leave land sheep-free a month
Lamb Dysentery	Acute diarrhoea, usually fatal Affects new lambs	Most shepherds inject ewes before lambing or lambs at birth. New vaccines protect against several diseases
Pulpy Kidney	Good lambs die suddenly Hold head in abnormal position, go into coma. Caused by toxins produced by soil bacteria	Ewes injected before lambing and lambs at 1—2 weeks
Pregnancy Toxaemia (Twin Lamb Disease)	Metabolic upset in pregnant ewes Walks unsteadily and may die. Often but not always associated with low energy diet. Caused by toxins released when body fat is suddenly utilised	Good food in late pregnancy Care in changing diet
Braxy	Animal moves jerkily and dies in a few hours Body blown up with gas. Can be caused by sudden chill or sudden change of diet	Vaccine can be used on remaining lambs, which should be moved to fresh ground

Animal	Normal Pulse rate	Normal Temperature oF	Normal rate of Breathing Respirations per minute	Gestation Period
Cattle	50–60 (cow)	101–102.5	12–16 (cow)	284 days
Sheep	75–80	102–105.5	12–20	147 days
Pig	70–80	101.5–103	10–16	115 days
Horse	38–42	100–101	8–12	340 days
Goat	70–80	102.5	22–26	152 days
Rabbit			40–60	30 days
Fowl		105–106		

Periods of Season	Duration	Return after Parturition	Periodical Return
Mare	5–7 days	7–10 days	2–3 weeks
Cow	1–2 days	21–28 days	3–4 weeks
Ewe	2–3 days	4–6 months	17–20 days
Sow	2–4 days	3–4 weeks	20–21 days

12 growing crops for animals

Crops in great detail are outside the scope of this book, but we must consider them in relation to our overall plan, for the provision of animal feed. Foods useful for animals will often be good for us as well. Wheat is the dream of every self-sufficiency enthusiast, but perhaps swedes are more realistic.

Soil. To talk about crops we must begin with the soil, which as we have seen, varies widely. The fertility of a soil, or the way it grows crops and feeds animals, is partly due to natural fertility, the result of its mineral content, climate, drainage and slope. Thus you will hear old farmers referring to individual fields as being mysteriously good for certain crops or stock. The actual condition of the soil can fluctuate, depending on how it is treated; it can acquire or lose fertility and structure.

The amount of organic matter in the soil is very important. This is decaying plant or animal life, which improves the physical structure of the soil and as it decays, also releases plant nutrients. We can also alter the drainage and the acidity, and our cultivations, crops and the weeds will all have their effect. These things are under our control.

The soil chart earlier outlines the main soil types and these will determine to a certain extent the sort of crops which can be grown.

Lime. Some soils are naturally acid and this will be apparent when acid weeds such as sour dock begin to show. Many organic farmers who have built up a good soil structure no longer need to apply lime; but it is very often needed when land is being reclaimed or improved. Ground limestone is a

natural form which can be spread on acid soils to correct the pH and to ensure a better breakdown of humus into plant food. But the lime should not be put on at the same time as manure, or the breakdown will be too rapid.

You will remember that pH is the chemists' way of expressing the acidity of a soil; the neutral point is 7 and acid soils are below this on the scale, alkaline ones above. Most garden crops prefer a soil just slightly the acid side of neutral but crops vary in their lime requirements.

Rye is one acid tolerant crop; its preferred pH is about 5.5. Oats and potatoes, grown in many acid soils, like a pH of about 5.0. At the other end of the scale, barley and lucerne like lime, and prefer a neutral soil.

Unless you are on a limey soil you may need to correct the pH initially and then to top up occasionally. Lime is lost every year by the uptake of crops and stock and drainage. It has been calculated, for example, that a beef animal removes about 35 lb of calcium from the land in its bones!

Materials used for liming

Material	Chemical Formula	Remarks
Ground limestone	calcium carbonate $CaCO_3$	most usual form
Burnt lime (quicklime)	calcium oxide CaO	more concentrated, better for carrying distances
Hydrated lime (slaked lime)	$Ca(OH)_2$	used in gardens, too expensive for fields
Magnesian limestone	$MgCO_3$ and $CaCO_3$	also supplies magnesium

(1 tonne of burnt lime equals 1.83 tonnes ground limestone)

Rates of Application. From about one ton to ten tons to the acre, depending on how acid the soil is; firms which sell lime will test the soil and are fairly trustworthy in their recommendations. Once the acidity has been corrected, a ton to the acre every four years or so should replace losses. This is the conventional view; some organic farmers would disagree and say that their soils, once balanced, can maintain their own lime status.

These are the functions of lime:
— speeds up the decay of humus
— promotes nitrification-making nitrogen available to plants
— helps nitrogen fixing in legumes
— removes sourness in soil
— makes insoluble iron phosphates more available
— helps to make potash available
— opens up stiff clay soils
— has a binding effect on sandy soils

Fertilisers

These are applied to replace the losses of plant nutrients in the soil; in modern farming the idea tends to be to 'feed' one crop at a time, but those with a feeling for soil will have a long term policy and will think first of feeding the soil. With the soil in good heart, the crops should be healthy.

Plant nutrition is an involved subject, which has tended to be over-simplified in the past. It is generally assumed that the

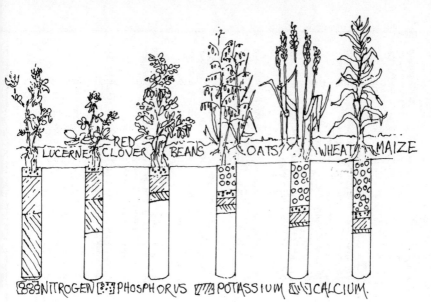

LUCERNE RED CLOVER BEANS OATS WHEAT MAIZE

⠿ NITROGEN ⠿ PHOSPHORUS ▨ POTASSIUM ⬚ CALCIUM.

What Crops Remove from the Soil

three essentials for plant growth are nitrogen, phosphorus and potash, plus of course, lime. Other elements are needed in small amounts. But organic farmers believe that feeding the plant by large applications of simple chemicals short-circuits the biological cycle, and produces crops which are less healthy and resistant to disease. They feel that these chemicals also cause loss of soil structure and contribute in this way to erosion. The association of certain fungi with the roots of plants is thought to be a way in which some plants gain food, and this kind of delicate balance is upset by chemical fertilisers. But let us consider the foods in turn.

When a crop is deficient in nitrogen, leaf growth is poor and young leaves are pale green, older leaves purple and red. Legumes, the pea and bean family, of course do not need to be given nitrogen because they carry on their roots bacteria capable of fixing the nitrogen of the atmosphere. These crops can thus leave nitrogen in the soil for the benefit of subsequent crops, and they play a useful part in a rotation. In gardens they come before brassicas, in the field often before wheat. Otherwise, nitrogen in the natural state is obtained by the plant from decaying manure.

Phosphorus deficiency is sometimes seen in young cereals, which are bluish purple with dead tips to the leaves. In border-

ORGANIC FERTILISERS

Fertiliser	Description	Uses
Farmyard Manure	Value depends on kind of animal Dung from fattening animals more valuable than from breeding stock; also varies according to food fed	Valuable for humus and nutrients it contains 16 tons to the acre for potatoes will also benefit subsequent crops: or put it on kale 10 tons contains about 100 units N 60 units P_2O_5 and 100 units k_2O
Liquid Manure (Slurry)	Can be a problem; smells and pollution of water courses. A lot is now produced by intensive livestock units	Useful for giving moisture to compost mixtures Better as an activant than applied straight to the land
Green Manures	Crops grown to plough in. White mustard most common, sown at 8 to 15 lbs per acre and ploughed in after 6 to 8 weeks. Lupins sometimes used — they fix N	Most useful on light soils. If mustard used, will also 'clean' the land of animal worms. If vetches used, add N. Does not increase humus much
Seaweed	Traditional manure in coastal areas. Organic matter quickly breaks down. Can be stacked in piles for use later, mixed with manure. Sometimes burned — residue a potash fertiliser	10 tons contains 100 units N, 20 units P_2O_5, 280 units k_2O. Also 360lb salt, plus iodine. Spread on land in Autumn — up to 45 tons per acre in Jersey!
Wastes usually too expensive except for gardening	Dried blood 10 - 13% N Meat and Bone Meal Hoof and Horn ground 13% N Shoddy (waste cotton and wool) 15% N	Variable composition, used in market gardening Nitrogen lasts longer than chemical fertiliser 100–300 units of N per tonne Applied 1–2 tons per acre
Ley	After several years build up of fertility, grassland is ploughed. Animal manure will be included *But* seedbed may be difficult and wireworm population may be high	Humus valuable for soil; the fertility released will grow several years' cable crops
Compost	Waste material — food scraps, manure, weeds etc., piled up, allowed to heat and converted by the action of bacteria into crumbly, odourless humus. Needs nitrogen activator to get it going	Probably the best plant food

CHEMICAL FERTILISERS

Fertiliser	Description	Uses (sold 'straight' or as compounds of two or more)
Nitrogen		
1) Ammonium nitrate $NH_4 NO_3$ (about 34% N)	Sold in granules in sealed polythene bags	Promote leaf growth generally. Make bombs! Make grass grow faster. (Also makes soil more acid)
2) Ammonium nitrate lime (21–26% N)	mixed with lime so as not to cause acidity. Absorbs moisture, so careful storage needed	Top-dressing grassland
3) Urea (45% N)	most concentrated, but expensive. May reduce seed germination	Mainly for grassland
4) Sulphate of Ammonia $(NH_4)_2 SO_4$	Not so popular as it was. White crystals by-product of gasworks	Top dressing
Liquid ammonia (Anhydrous ammonia etc)	sprayed or injected into soil	Grassland
Phosphates (expressed as a percentage of phosphorus pentoxide (P_2O_5) equivalent).		
Ground rock phosphate 29% P_2O_5 (insoluble)	Natural rock, ground to a powder	on acid soils with high rainfall. For grass and cabbage type crops
Superphosphate 18–21% P_2O_5 (soluble)	Ground rock, treated with sulphuric acid. Also contains gypsum ($CaSO_4$)	suitable for all conditions
Triple superphosphate 47% P_2O_5 (soluble)	Ground rock, treated with phosphoric acid	
Basic Slag 8–22% P_2O_5	By product of steel manufacture. Poisonous to cattle, so see it is washed off leaves before they graze	Best results on acid soils
Potash (expressed as a percentage of potassium oxide (k_2O) equivalent).		
Muriate of Potash (60% k_2O)	Does not store very well, and hard to spread, but most usual for farm use	where potash needed on farm. (e.g. potato crops)
Sulphate of Potash (48–50% k_2O)	made from muriate of potash, more expensive	used for market garden crops
Kainit and potash salts (12–30% k_2O)	Mixture of potassium and sodium or magnesium salts	Good for sugar beet

161

line cases, a deficiency may only show if the weather turns warm and wet. Potato tops may be spindly and root crops very small.

Potash deficiency in cereals shows up most if the spring is cold and dry. The tips of the leaves turn yellow. Potatoes need plenty of potash, and the leaves show a deficiency by having dark green foliage in early summer and later turning purplish bronze. Legumes get yellow spots, which turn brown.

These plant foods can be replaced by organic or chemical fertilisers, and the tables summarise the best known of these. In general, the organic ones tend to improve the soil structure as well, and they enhance the soil life of insects and micro-organisms. They tend to have a long-term effect, which may not be felt immediately on application.

Some of the chemicals are naturally occurring ones, such as ground rock phosphate, and should do little harm in moderate quantities, but organic farmers fear the effects of using chemicals on the soil life. Chemical fertilisers are expensive and backyarders will want to use them sparingly in any case, and to make the best use of home produced manure and compost. We are after all more concerned with quality than with size of yield. The other reason for doing without chemicals is that they represent in excess a reckless use of limited resources, imported often from faraway places. They say that over three tons of coal or their equivalent are needed to produce one ton of nitrogen. Yet only about half the nitrogen applied is actually taken up by the plant; the rest is washed out, often to cause trouble in rivers and water courses by encouraging immoderate growth of weed. The problem of excess nitrates is creeping into all kinds of places.

To get the best out of organic materials, they should really be composted; see the works of Sir Albert Howard and others on this. Sir Albert was concerned in the first place with restoring the impoverished soils of India, and to this end he studied the Chinese methods, which have kept many of the soils of China fertile for four thousand years. The principle is to return all wastes to the soil; and the way to render them harmless is to make compost.

On a large scale, compost can be made by aerating manure heaps, poking them with hedging stakes and turning them with machinery to encourage the development of the right sort of bacteria. On a garden scale it can if necessary be turned by

hand; the Soil Association and the HDRA have much literature on the subject. Compost makers say that they save time,
because when raw manure is put onto the soil it takes time
for it to break down sufficiently to be available to plants. In
fact some materials such as straw will actually detract from
the available plant nutrients while decaying. But compost is
all ready for use, and the heat produced by the bacteria will
have killed any weed seeds or pathogenic organisms.

Rotations

This means the organised succession of crops in a field, for the
good of the land and the crops; rather an old-fashioned con-
cept now, when continuous corn growing is the accepted practice. However, organic farmers make use of rotations because
monoculture is not to them acceptable. It is certainly contrary
to the arrangements of nature, which always has variety of
animals and plants, depending on each other in an ecological
system.

Control of weeds is one advantage of a rotation, replacing
the use of sprays. The root crops are referred to as cleaning
crops, but you have to do the cleaning! When there were rows
of roots in a field the practice was to get between the rows
with a hoe and knock out all the weeds, several times in a
summer, which discouraged them for a while. These days the
corn crop is regarded as a cleaning crop because it gets more
sprays than other crops.

Disease control is another benefit of a rotation. Most crops
become susceptible to attack if grown too long on one patch
of soil. The ones to watch are potatoes, peas and oats; they are
likely to be attacked by eelworms and nematodes. Commercial farmers reckon to grow potatoes in the same field for no
more than two years in eight, and they will not be consecutive
years. I have not made a long list of plant diseases because with
healthy soil and good rotation they should not be a grave
problem.

Rotations should be good for fertility, because different
crops will take out of the land different nutrients in differing
amounts; so it will be less exhausting if a variety of crops is
grown. Cereals make heavy demands, whereas grass stores up
fertility, partly because of the dung of the animals grazed on

it and partly from the mass of root material. This is the basis of ley farming. A ley is a temporary grass break of a few years as part of a rotation.

Another point to remember is that crops like potatoes can take a great deal of manure if it is available; some of this will be there to benefit the next crop. The old restorer of fertility was the folded crop such as turnips; sheep were folded or confined on a small area at a time and their manure was left behind. In those days they even fed rich oil cakes to the sheep, just for the added benefit in the manure, but this would now be far too expensive. However, it is a point in favour of buying in concentrates for your animals. If you feel that this is less than self-sufficient, remember that what the animals do not use goes back to the soil.

The famous Norfolk four-course rotation was usually wheat, roots, barley, clover (actually one year grass and clover ley). Then wheat came round again and benefited from the nitrogen left behind by the clover. Under roots the land was cleaned and received a heavy dressing of muck; and then another corn crop, barley, used the residue of this.

The Norfolk Four Course Rotation

A more modern rotation on a big farm might be corn, roots, corn, ley, ley, ley (a three year ley to build up the soil). Unless you settle for permanent grassland, you may be able to adapt this to the needs of your stock and land.

Garden crops are so well covered in organic gardening books that I thought it best to concentrate here on some crops for animals; this does not necessarily mean that you will need vast acres on which to grow them. They can all, or nearly all, be grown in the garden and if you are not sure how well a particular crop will suit your conditions, it might be a good idea to grow it on a garden scale first, for a trial run. The only limitation to this might be excessive bird damage to a small area of cereals. Some crops such as comfrey might only be needed on a garden scale. Many of the garden foods grown for human consumption will also provide useful fodder of course; spare cabbages, outside leaves and garden weeds will all be eaten up with relish.

Cultivations

The object is to keep the soil in a fit condition for crops to grow in; and the first and most important requirement is a good seed bed. The seed needs the right conditions in which to germinate, and before it does so it must be protected from birds. So it has to be buried, but not too deep. Different seeds have different requirements, partly because they vary so much in size; small seeds will generally need a finer seed bed, which means knocking the soil crumbs down into a smaller size. Anyone with a garden will be familiar with this operation.

Starting with grassy or weedy land, pigs are the best line of attack. They will eat off the vegetation, turn over the soil and manure it at the same time. It would be worth hiring a neighbour's pigs for the purpose if you have none; it is so satisfying to do so many useful jobs at once, and the pigs enjoy it so much.

Without pigs, the vegetation can be eaten down as close as possible by sheep or goats. If the weeds are high and coarse, goats will clear them, and briars too. Sheep will graze down close to the soil. Then the land can be ploughed.

Turning over the land by plough or spade will be the next job; but if the pigs do it, all you will need to do is to level it up afterwards. Think of the work they save! Getting a neighbour to plough up a patch will be good experience if it is

your first encounter with this work; later, you could have a go yourself.

After ploughing, the tradition was to leave the land in furrows to let the frost break it down and kill pests. Another school of thought deplores the leaving of bare land, and many people feel it is best always to have a crop covering the ground as far as possible; in which case, you will if conditions are right, harrow down the furrows straight away and proceed with the next crop. The weather may dictate what you do and your procedure will vary from year to year according to the season.

The one rule is — don't prepare a seedbed until you want it: once the soil is knocked down into a fine tilth, it may set hard like concrete with the wrong weather, so leave the land fairly rough until sowing time and then try to get the seedbed right and the seed sown quickly. Once the crop has germinated and covered the ground, it will prevent the soil from becoming too impacted.

Harrows are used to break up the clods. Spike harrows will probably do after pigs, but ploughed land which has been in grass will need disc harrows to cut up the vegetation into small pieces so that it rots down quickly. After several workings with this, a lighter zig-zag harrow can be used to get the required fineness of tilth. The seedbed can be consolidated with a roller if necessary. Sometimes a Cambridge roll is taken over the field after sowing, to press the seeds lightly into the earth. Hand sown seeds are even more likely to be picked up by birds than those sown by the seed drill, which injects them through spouts into the ground.

The seedbed should be a little rougher in autumn than in spring, because in wet winter weather the soil may 'cap' or form a hard crust before the seeds come through; and also, a fairly large crumb may give a little shelter from frost to the new plants. One example of such an autumn sown crop is winter wheat, which can be put in from late September until early February, according to the weather. October is probably the best time. Winter wheat is useful because getting such a good start, it will be long and green by spring and in some cases it can be grazed off lightly by cattle or sheep, as a spring 'bite'. This does not harm the crop; in fact it encourages 'tillering', the production of more shoots. The other advantage of winter sown cereals on a big farm is that it spreads the work load more evenly; all the corn does not have to be sown at once in spring, and the harvesting is staggered a little because the winter corn will be ready first.

Spring seed should be sown when the ground is beginning to warm up after the winter; our lecturer told us to put our faces down to the earth, the better to feel what it would be like down there.

The seed can be put in by hand or machine; it is a fairly easy job by hand, even on a field scale. Corn and grass are usually sown in rows by a drill, but they need not be hoed like roots, so they can easily be sown broadcast. You do this by walking along towards a marker at the other side of the field, throwing out seed from a pouch to the left and right. A barrow drill is like a wheelbarrow and this will sow in rows. A fiddle drill is a refinement of the hand method, where you walk along fiddling away at the instrument which releases seed at an even rate. Or you could borrow a fertiliser spinner and throw the seed out with that. We have sown seed successfully by all these methods; the initial problem is working out the rate of flow of the seed for a given seed rate per acre. For hand seeding, a higher rate should be allowed.

The hoeing and singling out of roots is the next job before harvest, but this is traditionally done by wielding a hoe and has never been satisfactorily mechanised. The less said about hoeing the better; you just have to get on with it!

167

Comfrey. Think hard before you decide where to put this, because it is a perennial, there to stay. It produces large rather hairy leaves similar in shape to those of the foxglove and they make good fodder, high in protein. If you like medicinal herbs, this is essential; it has healing properties. Comfrey of one or two kinds grows wild in Britain, but not in America. The commercial kind, Russian comfrey, is a hybrid and does not often set seed; but crowns can be obtained from HDRA and other places.

Comfrey needs deep soil with plenty of manure in it. The roots go down a long way, and so when established it will resist drought. The yield is quite high about 40 to 60 tons to the acre over the growing season, obtained in several cuts. Comfrey can be used for up to a third of the pigs' ration; feed it wilted for best results. Pigs will dig it up, so it is best to cut it and take it to them rather than letting them graze it.

Perhaps I should mention briefly some of the other uses of this plant, about which Lawrence Hills has written a most interesting book:

— as a table vegetable; the young leaves are cooked like spinach, with the tough midribs removed. Health enthusiasts put the leaves in a liquidiser with water to make a drink. Comfrey supplies vitamin B12 which is rare in plants.

— Comfrey tea is an old herbal drink; the leaves are dried in summer in the sun. It can be mixed with Indian tea to make it more palatable and this mixture is a very smooth drink, taken for rheumatism by old Yorkshire folk.

The *allantoin* in comfrey apparently makes wounds heal and kills bacteria. Since it makes fortunes for no one it has not been allocated funds for research, but the HDRA is now initiating work on the subject.

— the high potash content makes the plant a useful green manure, particularly for potatoes and tomatoes. The leaves are wilted to prevent them growing again, and laid between the crop rows; or they can, if ready in time, be put in the bottom of the trench when potatoes are being planted. It can be made into a liquid manure for feeding tomatoes in pots; the comfrey leaves are steeped in water for a few weeks.

— the plant is very useful for treating scour in calves and piglets.

Kale. This is useful because it is one of the few providers of green leaves in northern winters. There are several varieties; some, for example, marrowstem kale, will probably grow too big for gardens — we have grown it to a height of five or six feet. It certainly provides bulk on a field scale, but the huge stems may be too tough for some animals. Marrow stem is not frost resistant, so it is eaten off by the end of December; some varieties like thousand head are better in late winter and by choosing varieties carefully you can find a succession of kale from the autumn right through until the spring. Cows may need extra phosphorus if they eat kale all winter.

Kale is easy to grow; sown in late spring either in rows or broadcast (scattered), it usually makes its own way without any hoeing or singling and it is resistant to many of the pests and diseases of the cabbage family. Curly kale will make green salads for the table and a winter vegetable, cooked in milk.

For a good crop of kale, put as much muck on the land as you can before ploughing. The yield could be from 15 tons per acre upwards. 7–8 lb kale replaces 1 lb pig meal. Cows eat about 50 lb per day and more for larger breeds.

Chicory. This is another deep rooted plant; you could grow it in the garden as a path edging, for rabbits and poultry. Chicory is rich in minerals and a tonic for all stock, especially rabbits. Young leaves will go into salads and the roots can be roasted to make a coffee substitute. It prefers a soil with plenty of lime; chicory is often sown in herbal leys along with the grass. Apart from adding to the mineral content of the hay and grass the chicory is beneficial because the roots break up

the soil and asist with drainage.

Turnips and Swedes. Both useful crops, they differ in that swedes have a higher dry matter content and so are a slightly more concentrated food; but turnips score by having a shorter growing season, so they can fit into a rotation almost as a 'catch' crop. One can hope for about 20 tons per acre, perhaps more of swedes. These crops like a cool moist climate and hate an acid soil; unless there is plenty of lime they will be in danger of club root, the scourge of the brassica family. Brassicas should not be grown for more than one year in the same place, so swedes should not follow cabbages in rotation.

Turnips have hairy leaves of a light green, growing straight out of the bulb. Swedes have a neck from which the leaves grow, and swede leaves are smooth and grey-green. Both have several uses; as a table vegetable, and for various classes of stock. Sheep can be folded on the crop, when the land will get the benefit of their manure — a ewe with a lamb could eat about 20 lb of swedes a day. Or they could be grown in an odd corner, and carted to the animals. They make good food for cows and goats, and rabbits like them. Pigs will appreciate them for part of the ration, but they are too bulky for a main pig feed.

On a field scale, the seed rate is about 6 lb per acre if the seed is scattered by hand. The plants are thinned out later. In

gardens they are grown in drills 1 ft apart and thinned to about
a foot between plants. The thinnings can be eaten by stock,
or as a green vegetable.

Some of the less hardy varieties should be lifted before severe frosts set in, and stored in a clamp. Turnips are often lifted
in October and swedes in November, but they would go on
growing after this in mild weather. Purple Top is the hardiest
swede.

Cabbages. These are useful stock feed and many of the varieties will also be there for human consumption. Cabbage likes
moist and heavy soil. The danger of club root is of course, to
be remembered when planning the rotation.

Another drawback with cabbages is that usually they have
to be sown in a nursery bed and planted out as seedlings, which
is fine on a garden scale, but tedious without mechanisation in
a field. Some farmers are now drilling the seed in the field with
wide spacings and thinning later as you do root crops, rather
than transplanting. In either case they should end up with 2 ft
space between plants.

There are early and later varieties and you can have cabbage
all the year round for stock feeding if this is planned. Most
people find them more useful at the time of year when there
is no grass, but on a very small acreage, as William Cobbett
points out, a diligent cabbage grower could make enough food
to keep his cow on a very small patch indeed. In his 'Cottage
Economy' he tells us exactly how to go about it, digging the
plot by hand. Cows can eat about 50 lb cabbages a day; pigs
can have them ad lib, but will eat more if they are cooked.

Mangolds. These are grown in mild climates with plenty of
sunshine. They need warmth, but otherwise are a remarkably
hardy plant, attacked by few pests and diseases. We have
grown them successfully in north Yorkshire, but the yield
would have been bigger with a little more sun. They are quite
suitable for sheep, cattle, goats, pigs and rabbits, but not for
us except for wine making.

Mangolds are not resistant to frost, so they have to be lifted
in early autumn and stored in a clamp. Cover with straw and
then with earth to keep out the frost. They are not ready for
feeding until ripe, which is after December, as they mysteriously ripen off in the clamp. My chief memory of mangold
growing is keeping the weeds down in early summer, crawling
long the rows on our hands and knees. We sowed them at a
rate of about 10 lb per acre in late April and singled them to

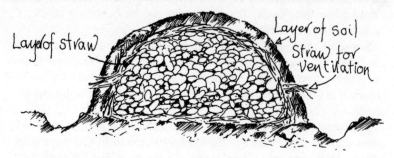

Layer of straw

Layer of soil

Straw for Ventilation

about 10 inches apart. The yield varies with the weather, but you can hope for about 20 tons per acre.

Fodder Beet is a cross between sugar beet and mangolds and has a high dry matter content, which makes it the most useful of the roots for pig feeding, 6 lb equals 1 lb pig meal. Another advantage of fodder beet is that it needs no ripening period and can be eaten when ready, in the autumn.

The tops have good feeding value as well, unlike those of mangolds. Sugar beet and turnip tops are all good stock feed, wilted first to rid them of the excess oxalic acid. In fact about 30 lb sugar or fodder beet tops equal 40 lb mangold in feeding value, so they are better food.

Living in a sugar beet growing district, we used to find odd beet on the roads and they made good rabbit food; but we have never grown them. The seed would be hard to get and home made sugar from beet is not practicable; better to produce honey.

Potatoes. These will be covered in any gardening book, but I mention them here because they are so useful for pig feeding. 4 lb of cooked potatoes will replace 1 lb of barley meal. Mashes for fattening poultry can also be made from potatoes, and they can be fed to cows. Fed raw, they are rather bitter and laxative but a cow can eat about 25 lb per day. A ewe may manage about 4 lb, while adult pigs can get up to as much as 14 lb of potatoes, cooked in this case.

Potatoes are a good reclaiming crop, a patch could be dug over by pigs, well manured and planted with potatoes; they could be grown on the ridge to keep them clear of weeds! They will perhaps be hoed while they are growing, and when you dig them up you are cleaning the land again, so by the time they are finished it will be ready for whatever crop you wish to plant next. (But not potatoes).

172

Maincrop varieties will be best when growing them on a
field scale, and earlies can be grown in the garden. Maincrop
are planted in April at a rate of 16 to 20 cwt per acre; they
should be ready for harvesting in September—October, and the
yield should be at least 10 tons per acre, possibly more. Commercial yields are up to 20 tons per acre.

Cereals. The table gives a summary of the various cereals
and their uses. Growing them on a small area is doubtful sense;
birds make such inroads into small patches of cereals and the
cutting and harvesting, not to mention threshing, by hand are
extremely hard work. The yield per acre is low compared with
other crops, which is why backyarders have not grown their
own corn very often in the past.

Our family have grown wheat and barley on a field scale,
harvested by our own ancient combine; it was good to eat our
own home grown bread, but it was not really got by the sweat
of our brows! One man who has harvested grain and threshed
it by hand is Sedley Sweeney, Soil Association member, whose
booklet, 'Smallholder Harvest' is a mine of information for
anyone who wishes to experiment with cereals. It will be wise
to choose a kind of cereal suited to your conditions; a wet
Welsh mountain side, for example, will not be the best place
to try wheat. But I have seen some magnificent rye growing
high up in Wales. Rye can be grown up to 1,200 ft above sea
level; it would make a good reclaiming crop where you were
trying to improve acid land.

Beans. Many newcomers to backyarding, faced with expensive cereals, wonder why more use is not made of home grown
beans. Surely we could all be self-sufficient for protein? One
of the problems is the low yield; only about 18 cwt per acre can
be expected and the crop takes up a patch of land for a long
time; winter beans are sown in October, spring beans in March,
but winter beans are not too successful in areas with the same
climate as North Britain. Soya bean meal is the most useful
for stock feeding but this crop is not suitable for Britain at all.
The beans we can grow in Britain are not so high in protein
as soya; 2 cwt of our beans equal 1 cwt of soya plus 1 cwt of
barley - but the quality of the protein from the bean is slightly
lower. Of the winter sown varieties, grown in the south of
Britain, one variety is Maris Beaver. These will be higher yielding than spring beans.

Of spring sown beans there are two kinds, tick beans e.g.
Maris Bead and the ones grown for stockfeed, horse beans e.g.

Cereal	When sown and seed rate per acre	Some of varieties	General Characteristics
Wheat hardy, likes dry climate	Winter 1-1½cwts Spring 1¼-1¾cwts (yield less)	*Winter* Maris Freeman, Flinor, Mega *Spring* Maris Dove Sappho	Deep rooted, best on very fertile soil Land must be well drained
Barley likes light land, plenty of lime, warm dry weather	Usually sown in Spring, but there are autumn varieties 1¼cwt	*Winter* Astrix, Senta *Spring* Julia, Zephyr, Proctor, Vada	Shallow rooted. Will 'lodge' if land too rich. Needs no manure if previous crop was manured
Oats likes acid soil cooler & wetter climates	*Winter* yield best *Spring* susceptible to Frit Fly 1½-2cwt	*Winter* Maris Osprey, Peniarth *Spring* Leanda Forward	Will grow on most soils but been replaced commercially by barley but oats more resistant to disease. Manganese deficiency if too much lime in soil
Rye will grow on poor acid soils in dry places	Usually winter For grazing, sow late Aug-early Sept	*Winter* (grain) King II Petkus, Dominant Grazing: Ovari, Rheidol	Very frost hardy but does not yield so well as other cereals on good land
Maize will not grow until temp 10°C needs sun for grain production	Spring only	Early Maize *Grain* Dekalb 202 *silage* LG11 *sweet corn* Early King *Late maize* Grain—Anjou 196 Silage Inra 321 Sweet corn October Gold	Dislikes wind, needs plenty of sun, rich deep soil but not too heavy. Resistant to many diseases but susceptible to bird damage
Mixed Corn (dredge corn)	Both winter and spring. Barley and oats a common mixture 1¾cwt sometimes cereals plus peas or beans mixture called maslin	Varieties chosen which ripen at the same time. *Or* could sow beans earlier than oats	Do better than one cereal by itself

Cultivations	Harvesting	Uses	Average yield per acre
Likes a firm seedbed but not too fine. May graze winter wheat late March if weather dry	Ready for binder when straw yellow grain cheesy Combine—10 days later. Straw—whitish grain hard and dry easily rubbed August usually	Bread Poultry feed Straw for thatching	Grain 32cwt Straw 24cwt
Winter barley— slightly lumpy seed bed. Spring barley seedbed loose on top and firm underneath	Winter barley early, July-Aug; bird attack a problem. Spring— Aug-Sept DO NOT cut until dead ripe ears turned down even for binder	Feeding livestock Malting for beer Straw for bedding fodder. Pearl Barley for soups. Barley kernel for puddings	Grain 28cwt Straw 24cwt
Similar to wheat	Cut with binder when straw still green, not so easy to combine as barley	Fodder for all classes of stock and people Oat straw is useful fodder but variable	Grain 24cwt Straw 1-2ton
Similar to wheat	Should be quite ripe when cut, or will be difficult to thresh	An early Spring bite for grazing animals. Grain for rye bread straw for thatching Can graze twice, or once + grain crop	Grain 20cwt Straw 30-40cwt
Plant 2 inches deep in level moist bed, end of April. Weeds kept down	For silage—end Sept. Grain. Oct-Nov. Frost helps dry grain. 15% moisture content for storage or can freeze	Forage crop (green) in cold climates Sweet corn—vegetable Poultry and pig food Part of cattle ration	
Sometimes grown after grassland is ploughed	When both or all are ready	Makes good stock feed—better protein when peas and beans included. Used to make maslin bread for human consumption	

Suffolk Red. Horse beans should be sown in early March at the rate of 2.5 cwt per acre. The width between rows varies and does not seem to affect the yield.

Beans do not like an acid soil; they can make their own nitrogen and their main need is potash. Manure or compost can be ploughed in when the ground is being prepared. Horse beans do well on heavy soil.

They are ready for combining when the pods are black and if harvested dry will store well. Beans used to be cut with the binder, stooked and left in the field for two or three weeks; in this case they were cut when the pods about halfway up the stem were black and the scar on each bean where it is attached to the pod was also black. The green beans would blacken in the stook. The 'straw' from binder-cut beans used to be fed to stock. The beans would be stacked and threshed after Christmas. Spring beans will be ready for harvesting in September or October.

Many beans will be shed from the pods when the crop is harvested; choose a dull day for the work, and preferably graze the field with sheep or some kind of stock straight afterwards, to pick up the fallen beans.

13 the grass crop

The title of this chapter is deliberate, to remind us that grass should always be treated as a crop. It is a most valuable crop - two thirds of the farmland in Britain is covered by it. It feeds most farm animals and at the same time builds up soil fertility. Grass is the vital link between animals and the earth; it forms an important part of the organic cycle. This is why it is no bad thing to have an all-grass holding. It is not monoculture; with a variety of animals it can be good organic farming, and the most sensible crop in a wet climate. Grass is the most important backyard crop.

What you can do with it depends on which of the three main types of grassland you have, and whether it is good or poor of its type. Owners of upland holdings often find themselves managing rough hill grazing. The mountain grasses are not very valuable to lowland eyes, since they produce little bulk, but mountain sheep like them and this is important if you keep sheep. The various fescues, bents, nardus and molinia grasses, plus heather and gorse, make up the sward or grass carpet. Often there are rocky outcrops, and the soil may be thin and acid. Improvement can be brought about, but reseeding is often impossible because of the steep slopes or lack of soil depth.

Other types of permanent pasture are variable in the extreme. The term means pastures which are never ploughed. Good ones will contain a high proportion of perennial ryegrass, plus useful herbs and inevitably some of the weed grasses as well. These pastures may remain unploughed because they are convenient for the buildings, or they may be wet or steep.

The third type of grassland is the ley, or temporary sward, sown down to grass for from one to four years as a break from other crops in a rotation. It has a good chance of producing well because the seeds mixture used will contain productive grasses and the weeds should not have had time to creep in. But this kind of grassland lacks the firm springy turf of old pastures and is therefore more liable to poach in wet weather.

Species

The plants making up the sward are grasses, legumes and herbs. A herb is really anything not in the first two categories, and many of the deep rooted plants such as plantain are valued for their mineral content. The legumes are species of clover, lucerne and sainfoin; like peas and beans in the same family, they fix atmospheric nitrogen in the soil, which benefits the grasses grown with them. They also provide a good 'bottom' to the sward, but they like warm weather and start growing rather late in the spring in cold temperate climates. Where the weather is warm enough to encourage them, as in parts of New Zealand, they form an important part of the swards. But with a high proportion of clovers there is the danger of bloat if an animal overeats.

There are over 130 grasses, but only about twenty of these have any real agricultural significance; the table gives an idea of which are the most useful, and why. These are the species found in good grassland all over the world. The cultivated grasses have been developed like the cereals for food production. They are high yielding and their leaf is palatable and digestible to the grazing animal. Their growing season is longer than that of uncultivated grasses. Once a grass flowers and then sets seed, it grows tougher and less nutritious. You may have noticed in your own fields that the wild grasses flower earlier than the cultivated ones; and their bulk is less. Agronomists like those at the plant breeding station at Aberystwyth can tell you the exact date on which to expect the flowering of various strains, and therefore the best time to graze or cut them. Within the varieties they have bred strains with useful characteristics for different purposes and the seed merchant will be able to advise you on which will be best for your land and intended use. For example, S23 is an Aberystwyth strain of

178

perennial ryegrass bred many years ago for spring and autumn grazing.

Grass seeds can be bought from your local feed merchant as well as from the seed specialists. They will have a catalogue and will make up a mixture to your requirements; farmers usually buy seeds ready mixed rather than shopping for a pound of this and that. The mixture is a kind of recipe, a list of varieties and the rate per acre. It is unwise to sow a field with just one species of grass, as well as being boring for the animals. A nice mixture is much better and will be an insurance against the failure of one species.

This is an example of an orthodox long term mixture:

Long Term Seeds Mixture

Perennial Ryegrass	S23	6 lb
.. ..	S321	8 lb
Timothy	S48	6 lb
Wild White Clover	S100	2 lb
	seeding rate	22 lb per acre

Many farmers feel that a long ley is not complete without some herbs, for minerals and soil drainage effects. Sam Mayall of the Soil Association suggests chicory and ribgrass, one pound of each, as the most useful ingredients. He says that yarrow, burnet and sheeps parsley are also useful but will not do well on all soils and may be choked out by strong grasses. They are however good on sheep pastures with finer grasses; or they could be sown in a herbal strip at one side of a field and mixed with legumes rather than grasses.

Reseeding

Your own choice of grasses will depend on the area, and also how long you want them to last. Perennial ryegrass, cocksfoot, timothy and meadow fescue are grasses for long leys. For the shorter term, Italian ryegrass and other short lived varieties will be chosen.

The seed can be sown either in spring or early autumn. They can be put in direct or with a cover crop (undersown). It used to be common practice to sow grass seeds with a spring corn

179

SOWING DOWN GRASSLAND

Species	Latin Name	Uses
Perennial Ryegrass	Lolium perenne	Long use — permanent or long ley. Good in spring and autumn in most parts of the world.
Italian Ryegrass	Lolium multiflorum	Short-term leys. Early bite in spring.
Timothy	Phleum pratense	Long term leys. Pasture and hay, sown in mixtures with meadow fescue.
Cocksfoot	Dactylis glomerata	Early hay. Upland pasture. Adds humus to soil.
Meadow fescue	Festuca pratensis	Grows well with timothy. Good for hay and grazing on moist, rich soils.
Tall fescue	Festuca arundinacea	Used to be thought one of valuable grasses can be grazed in winter.
Red fescue	Festuca rubra	Poultry — Fine. Useful on poor uplands.
Crested dogstail	Cynosurus cristatus	Old-fashioned grass for sheep. Fills up bottom of sward.
Smooth-stalked meadow	Poa pratensis	Poultry and rabbits. Important pasture and hay grass in N. America.
Rough stalked meadow grass	Poa trivialis	Can be useful in longer leys. Good on rich soils in sheltered places. Winter grazing.
White clover	Trifolium repens	Usually included in long leys. Keep down weeds by creeping growth.

Advantages	Disadvantages
Palatable — the most useful grass, good for grazing and hay. Makes good turf.	Not so good in dry areas. Lack of production July—August.
Very palatable. Green in winter. Good bulk.	Dies out after a year or so, an annual plant.
Tolerates wet soil in winter. Very palatable.	Not good in dry areas or on light soils — shallow roots. Not very persistent, tends to die out in time.
Deep rooted and drought resistant. Good on poor soils. Good under trees.	Burns in winter. Can grow coarse and tough, not so digestible.
Good yield. Fairly early in spring. High digestibility	Does not establish well in mixtures with cocksfoot and ryegrass.
Resists drought. Winter hardy.	Some kinds are coarse. Less palatable than meadow fescue.
Good on poor shallow land.	Not high yielding.
Resists drought.	Wiry and not very productive. Expensive.
Fine. Good on dry soils.	Less productive.
Covers the ground well. Can stand wet. Very palatable.	Dies out in dry conditions. Low productivity.
All clovers fix nitrogen in soil. High digestibility. Long-lived.	No growth in cold weather. Too much clover can cause bloat in some circumstances.

Species	Latin Name	Uses
Red clover	Trifolium pratense	Late and early flowering varieties. Usually put in mixtures.
Alsike clover	Trifolium hybridum	Usually included in mixture in case red clover fails.
Lucerne	Medicago sativa	Usually main ingredient and used for cutting rather than grazing.
Sainfoin	Onobrychis sativa	Gives bulk to mixture and variety. Sow with cocksfoot for hay.
Yarrow	Achillea millefolium	In mixture gives variety and good root system.
Chicory	Cichorium intybus	About 1 lb included in mixture to improve health of stock.
Plantain	Plantago lanceolata	1 lb in mixture of herbal ley.
Burnet	Sanguisurba officinalis	Included in leys for sheep. Gives nice taste to butter!

Advantages	Disadvantages
Good hay yields.	Short lived.
Tolerates damp and acidity. Persistent.	Not so productive as red clover.
Heavy yields for hay and silage Very good for stock health. Useful on dry soils.	No good on acid soils. Not good for grazing, not very palatable or digestible.
Good on chalky soils.	No good on acid soils. Not good for grazing. Not so productive as lucerne.
Good for sheep Good in drought.	May be smothered by grasses.
Deep-rooted. Brings up minerals.	Difficult to eradicate if you want to grow other crops.
Minerals good for stock. Will grow anywhere.	Hay difficult to dry if too much in mixture.
Cures scour. Good for sheep and cattle.	Expensive.

crop. The corn shelters the grass seedlings, and when it is harvested, the grass is already established; I have seen under-sown grass give a good autumn bite. There can be a drawback; in a wet season the grass may grow long enough to hamper straw drying, although if the straw can be dried and used for winter fodder it will be a useful feed, being partly hay.

For this the grass seed is usually sown at the same time as the corn.

For direct sowing the seed bed must be just right, fine be-cause the seeds are small, but firm to give the roots a hold. A dry period may make germination patchy and you could have an anxious time. But the advantage of spring sown grass is that it can be lightly grazed in the summer, which will do it good. Young cattle or sheep are best for this first grazing; put them on for just a few days and then remove them. Try to do this in dry weather because hooves can do great damage to a new sward in the wet. A light grazing will encourage the growth of new shoots. Sometimes 2 lb or so of rape is sown with the grass seed as a very light cover. This can be grazed by sheep 6 to 10 weeks after sowing.

We have occasionally found it necessary to mow a new ley,

to cut down the annual weeds which were obstructing the narrower grasses. But unless it is being smothered in this way, new grass should not be cut in its first year; grazing will do much more good.

Autumn Sowing

The soil is warm after the summer and the weather may be damp, so early autumn is also a good time to sow grass seeds. The disadvantage is that clovers cannot stand cold weather and may be killed by early frosts before they are properly established.

Grass seeds should not be sown too deep. They can be broadcast by hand or drilled with a barrow or fiddle drill, both hand controlled implements. You may be able to borrow one from a neighbour. If there are several acres to sow, it may be as well to get someone to put the crop in with a seed drill at the right depth.

Improvement

Where reseeding is not possible grassland can be improved without it. If the grass is thin, with no bottom, more seed can be scattered on the field in the spring, and harrowed in. A good dressing of manure in the winter before this will help.

Where the old grasses have grown coarse and tussocky, or

formed an acid mat that does not decompose properly, heavy grazing with sheep will be the first stage of improvement. In upland areas it is quite usual to take in other people's sheep for the winter, or even for a few weeks to relieve pressure on overworked grassland. So a few imported sheep would help with the pasture improvement. After the grass has been eaten down, a severe harrowing will further break up the mat and aerate the roots. A rest, followed by good management in future, will improve the pasture a great deal. Fairly heavy grazing followed by a rest is better than continuous light grazing.

But be warned; continuous grazing with sheep alone will in the end lower the quality of a pasture, so mixed stocking is best. Introduce cattle if you can. It seems that the Scottish Highlands are poorer now since they were cleared of people to make way for sheep. When the crofters had the land, they had mixed stock including cattle.

There may be other problems with upland pastures. Bracken for example; dreadful stuff to eliminate, but worth it because it often grows on reasonably good land. If it can be cut regularly twice a year, in the end it will be discouraged; I have seen the spectacular results of Welsh experiments on this technique, where steep hillsides were cut (with student labour) by hand until the bracken began to disappear.

Dried, it makes good bedding, but do not be tempted ever to feed bracken to stock. It is true that sheep will nibble it often with no ill effects, but there is the possibility of bracken poisoning, and I have seen references to the discovery of a cancer-producing substance in the plant.

Heather is often burned off in the early spring; this encourages the growth of new heather shoots which make a feed for sheep. If the land will support more productive plants than heather, it can be discouraged by grazing the area heavily and then spreading lime. This treatment will also discourage moss.

Gorse used to be put through a chaff cutter and fed to stock in winter, but its feeding value is very low and no stock will touch it while it grows, so its removal will improve the pasture. The only place in which it should perhaps be left is in an exposed field where it could provide the only shelter from the weather; it does make a good winter protection for sheep. The easiest way to get rid of gorse is to allow pigs to root it up, but otherwise it is burned off in high summer when it becomes very dry, and this can be hazardous.

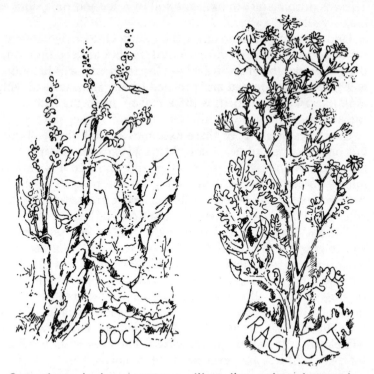

DOCK

RAGWORT

Sometimes docks or ragwort will spoil an otherwise good pasture and it is laborious, but best, to take them out by hand. Annual weeds such as chickweed are discouraged by a cut for hay and in fact may provide food in this way. Nettles look rather unsightly in the field, but provide nutritious hay and have been grown for the purpose in Scandinavia; the formic acid which causes the sting disappears when the plant is dried. Weeds are kept down when grassland is used to the full. If there are only a few animals, as may happen in backyarding, the paddocks should be correspondingly smaller so that they are grazed off quickly and then left to rest. A balance must always be struck between the needs of the grass and the needs of the animals; the real necessity is to avoid undergrazing or overgrazing.

Overgrazing occurs when there are too many animals on the area for too long. The grass gets eaten down to the roots, bare patches appear. Parasitic worm larvae build up and the stock does not thrive. If the grass never gets a chance to recover, it will not yield as it should. In upland areas where soil is thin, these are the conditions when erosion is likely after heavy rain.

There is nothing left to hold the soil in place and on a slope it is easily washed away.

Undergrazing is also bad for the sward. One or two animals set to graze a large expanse, will not be able to keep up with the growth of the grass. It will get long and coarse. The grasses will flower and set seed and proceed to die, and the field will look like an unmade bed. A thick mat of dead grass will choke the next season's growth.

Examples of either of these extremes are not hard to find; but how are they to be avoided? It is not always possible to have exactly the right number of grazing animals; it is impossible to predict what the growth of grass will be each season. And in late spring and early summer, more grass grows than will be available later, when the rate slows down.

The best answer is to divide up the grazing into sections and to graze each paddock a few days at a time. In a season when there is too much grass for the stock, one part can be cut and conserved as hay or silage. A light topping with a scythe or grass mower will remove the surplus where there has been undergrazing, with clumps of coarse grass left uneaten. Once it has been cut it can be left and the field grazed; the animals will eat it quite happily when wilted. The mower should be set high; pastures are never treated like lawns.

Overgrazing can be a problem where you are carrying a lot of stock and if it is a temporary problem the answer is to put the surplus animals into yards. There they can get fresh air and

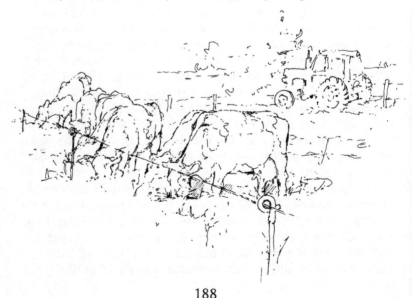

exercise, but the land will get a rest. If it seems a longer term problem, it will be better to cut down on stock than to continue to overgraze the land; but consider the roadside and wild foods first, as many a hardpressed homesteader has done.

Grazing systems vary according to many things, including fashion. But the paddock system is the best for the backyarder and I make no apology for mentioning it so often.

Hedges and ditches should be kept in order to help the grassland. To keep in stock and to shelter wildlife, hedges should be thick, and regular trimming will keep them so. If they are neglected, hedges grow straggly and thin, and they also grow out into the field, which would of course eventually go back to scrub. Ditches, if neglected, will flood the land and take away valuable topsoil. Attend to hedges and ditches once in every three years.

Grass Conservation

The amount of grass available for conservation will naturally vary from year to year. On a small place it will be better to buy in hay, but as we have just seen, hay making can be a help in grassland management. Do not feel guilty about buying hay; the extra stock thus carried will improve the fertility of your land.

Silage.

The process of ensilage is that of preserving a crop in a silo. Anything can be ensiled that will work up the necessary fermentation; I have helped to make silage which included potatoes and brewers' grains. Grass ferments very well. The carbohydrates in the plant cells are converted by bacteria present in the grass into acids. They are mainly lactic, acetic and butyric acid, so in the end the grass is pickled in its own juice. You can tell by the smell what sort of fermentation is going on; good silage smells pleasant and is mainly lactic acid. Up to 2% of the silage can be lactic acid. With a colder fermentation there will be a lot of butyric acid, which smells dreadful and makes a less palatable silage.

To encourage the lactic acid bacteria, heat must build up and air must be kept out; the grass is consolidated, often by driving a tractor over it. The grass is put into the pit, clamp or stack

189

either fresh or when slightly wilted, in which case the dry matter of the silage will be higher. The temperature should get up to about 90 degrees F.

Grass for silage may be cut at any time, but the feeding value of the silage will be best when the grass is cut at its best, just before it flowers. Later cut grass will contain plenty of carbohydrate to ensure fermentation; but if the grass is cut when very juicy and green, the protein will be high and extra carbohydrates may be needed. They can be added in the form of molasses; 1½ gallon per ton of grass is sprinkled onto the grass as it is stacked, after being diluted with an equal volume of water. About twice as much molasses will be needed for lucerne or any other legume. When this was my job, I did it with a watering can with the sprinkler attached.

Farm silage making succeeds because enough grass is put in at one time to get the necessary weight. Fermentation only proceeds if air is excluded. Backyards may not have enough material, or a tractor to drive over the heap to consolidate it — a trick that has its dangers, as you may imagine.

There is a silage method ideal for preserving small amounts of grass or any other material. Wilt the grass for a day after cutting, then pack it tightly into polythene sacks (usually obtainable free from farmers). Tie the neck tightly and stand in a heap. Airtight conditions ensure good preservation and the silage will be easily manageable. Once you have the trick of it, you need never refuse any unexpected windfall! We once did this with pig food, swill from canteens. After it came out of the sacks it was boiled, so that fermenting material was not fed to the pigs; but it kept very well like this for several months.

You can make a silage stack in the yard. We did this on a farm where I once worked, cutting lucerne about four times in the season and making four stacks. We watered it with molasses and jumped up and down on it; mercifully there was an elevator to take up the greenery when the stack got high. After it was finished we topped off the stack with ground limestone and then cow muck, to give weight and keep out the weather. The next year, this material from the top was spread on the land. It was rather a wasteful silage method because where the sides were exposed to the air, the silage was spoiled. It has been worked out since that wastage from a stack can be as much as 40% which is not acceptable. But the middle was lovely stuff for the cows, and the sheep ate most of the rest.

Feeding Silage

It is a variable commodity — but then, so is hay! Very generally then silage may have a dry matter content of 20%, up to 25% (the rest of course is water). It weighs 45 to 50 lb per cubic foot. 3 lb silage equals 1 lb hay in feeding value.

Cows; a small cow can take up to about 50 lb of silage a day plus a few pounds of hay.

Calves; good quality silage can be offered to calves when they are a month old. By 6 months they can be eating up to 15 lb a day.

Sheep; up to 12 lb per head for ewes in bad weather. 6—7 lb silage equals 1 lb concentrates.

Goats; they will eat silage once they get used to it, but noses will be turned up at first and they will never eat the poorer quality stuff. Quantities are similar to those for sheep, but the goats will decide.

Hay

Hay making needs a few days of good weather; therein lies the gamble. But it can be made in any quantity and at any level from the simple to the sophisticated. The idea is to cut for hay when the grass is right *and* there is a good weather forecast; in most places the nearest aircraft station will give a local forecast, which can be very helpful.

The grass should be cut for hay when there is a respectable bulk of crop, but before the grasses go to seed and become too tough. This will depend on what kind of grasses are present; poorer grasses and the upland kinds tend to set seed very early in the summer and never make much bulk. On the other hand, clovers keep their protein levels high longer than grasses and so will help to improve the quality of the hay.

The job of haymaking involves reducing the moisture content of the grass from about 80% down to 20%, when it will be safe to store. Some farmers use barn drying techniques, but most still depend on the sun and the wind. So good hay is rather a matter of luck, and making the most of opportunities.

It is important not to overcook your hay. If it is bleached to the colour of straw, you will have lost vitamins A and C and if it is too brittle the leaves will break off and may be lost before the bundle reaches the animal. But beware; if the hay is carted too soon and stacked when damp, it will heat up and cook in the stack. It will change from the lovely dark green of new hay to a brownish colour, and may have lost about a quarter of its feeding value. Severe overheating of damp hay can lead to combustion and this is the cause of some farmyard fires.

Early hay, made from young grass, is high in protein but the water content is also high, so it takes a lot of drying. Later in the season it will be easier to dry but less nutritious. It is not wise to try to make hay too late in the season, because the late summer days are not long enough to dry the grass, and the dews are wet in the morning. Sometimes a new growth in late summer can look very green, but the scientist tells us not to over-estimate its value; autumn grass is not so valuable as it looks.

Cutting Hay

The grass is cut in swathes, either with a scythe or by machine; traditionally, you start one swathe in (ie not at the extreme outside edge of the field) and go round in a clockwise direction, finishing in the middle. The hedgeback swathe is cut last, and in the opposite direction. For years we cut our hay with a horse mower converted to tractor use; horse machinery needs a man on the machine as well as one on the tractor. If a contractor cuts your hay he will probably use a big modern rotary mower which will do the job very quickly but needs plenty of room to turn.

Let the hay dry out on top, then turn it to expose the wet underneath. After this it is shaken out to let the air get to it. This is called 'tedding'. There are many turners and tedders with poetic names - we have an Acrobat and a Cock Pheasant helping us at haytime. You can buy them in farm sales and every farmer seems to have a different favourite. But for these you will need a tractor or a horse.

Do not despise hand tools. Hay can be turned efficiently by using a hay fork and wearing a large hat, provided there are enough people and forks. Rake it up into high windrows if you want the wind to dry it; scatter it out on the field if you are relying on the sun. It has to be worked until the moisture is out of it, and this depends on the weather.

In poor seasons or an uncertain climate it can be made into piles called haycocks in the fields when half dry, to preserve

it from showers if rain is threatened. After the rain, it can be shaken out to dry again. Or you can do as the Swiss do and hang it on tripods to dry. This is a good idea; the tripods keep the grass off the ground and let the air through it, and all except the outside layer is protected from the rain. The Soil Association booklet tells you how to make them. Tripods can be six or seven feet high and you will need about ten for an acre of hay.

In Wales we were often visited by swift showers coming in from the sea and a few hastily made haycocks often preserved our crop. In windy weather we had to weight them down with ropes and stones.

When we farmed up to twenty acres we made loose hay, and beautiful fragrant stuff it was; over this acreage, as we expanded in order to make a living from our land, we had to bale the hay.

Ideally hay should be baled when it would be ready to cart home loose, but in poor weather it is safer in bales and sometimes we have to bale when it is not quite dry enough, and leave it out in the fields in stacks. It is safer so than loose, ie protected from the rain to some extent, and if it rains a lot we have to cover the stacks with polythene sacks. But really wet hay will not bale; it won't go through the baler and can easily break the machinery.

Baling does make the hay easier to store and if there is room in your field for a baler to manoevre it will be simple enough to get a contractor to do it. This will cost about 15p per bale. Small quantities of hay can be 'baled' by hand, compressing the hay into a wooden box and tying it up.

14 the wild harvest

There is still a lot of wilderness left in the world, thank goodness; we must use it with care and not plunder it. With common sense, we can conserve and still forage for wild food for ourselves and our animals.

Do not get arrested for taking things from private land! The legal aspects for Britain will be dealt with later, but briefly, you can take wild things but not cultivated crops. Flowers, fruit and foliage may be picked for your own use, but not for sale. Trespass is a misdemeanour but not a crime. In fact, it should be possible to get permission to go onto private land. (And then there is the common land, the moors and wastes, shrunk now from their former vastness and the age-old peasants' stamping ground).

In general I think it is best to get permission to go into fields for anything more than a quick foray, unless it is common land. Then you can proceed with a clear conscience. Don't take a dog unless you are poaching, because animals may take exception to it and this could be bad for them or for you. A woman I knew once was attacked and injured by a newly-calved cow because she went near it with a dog.

It is now illegal to dig up plants; taking away a whole plant is in any case not quite the same as gathering leaves or fruit. We should aim to leave enough of the plant for it to flourish and it will be there to gather the next year. This is common sense; there are too many people now for us to be able to afford to be careless.

The animal kingdom is also beleaguered and in general perhaps we should not hunt them for food, with the exception of rabbits which are themselves a pest. (See the Backyard Book on Rabbits for ways of catching them). Fish are rapidly becoming scarcer — it may be better to grow your own.

In spite of the density of population in the places in the world where there is food for free, most tourists stick to routes and tend to keep to roads and lanes; you can soon get into the wild if you are willing to walk.

What can be found in the wild? First of all, food in abundance for our backyard animals. The grass verges of lanes are full of wasted opportunities. There are few people like the man I saw last summer, who was carefully raking up the cut dried grass at the roadside and loading it into his van. Naturally, these days there are hazards — traffic fumes pollute the edges near the road, dogs may foul anywhere. Some verges are sprayed with weedkiller by unenlightened authorities. A campaign to discourage this, called Save Our Wild Flowers, has had some success in Britain. As spraying gets more expensive, we can hope it will be given up. Who wants sterile roadsides?

Once I had a job which involved visiting farms, and for

years I drove round the countryside; and I studied the plants of the roadside. I know how valuable they are because when we kept a lot of rabbits in a suburban garden, these plants were our chief source of greenfood.

Rabbits, goats and poultry will all enjoy bunches of mixed greenery just as we enjoy mixed salads; the variety will give them a better chance of a balanced diet. Grass and clovers, vetches, dandelion, shepherd's purse and so on will make a good meal. Goats prefer to choose their own and if you have

to choose it for them, include a few branches if you can. Cows will graze very happily along a quiet lane; in Yorkshire this is a recognised practice, known as "tenting" where you wander along with the livestock, keeping an eye on it.

There are some poisonous plants, of course; see the table for some of them. Some of the plants which are poisonous when fresh are harmless in hay because the toxins are lost in the drying process, just as the nettle sting is lost. Learn to avoid the bad ones, but don't worry too much about picking them unknowingly in a bunch of greens; if they have choice most animals would leave them anyway.

What you can find for the animals will depend on the district; you may be lucky enough to get acorns for the pigs or comfrey for all the stock. Comfrey is about the best roadside bulk to be found. It is a matter of luck, but

if you live in the right place the wild greenfood can be as useful as another acre of land. Backyarders have relied on it to feed animals they could not otherwise have afforded to keep.

What is there in the wild for us? The table will give you some of the possibilities. There is something at every season of the year. If you happen to live near the sea, you are lucky; there is a rich harvest of seaweed, shellfish and drift-wood.

197

PLANTS TO AVOID

Plant	Usual Habitat and Appearance	Remarks
Yew (*Taxus baccata*)	Churchyards, old gardens. A dense evergreen tree.	All parts poisonous. Often kills farm animals in winter.
Marsh Marigold (*Caltna palustris*)	Wet boggy land. Yellow flowers in spring and early summer	Not poisonous in hay - toxins lost by drying. Irritating causes blisters.
Larkspur (*Delphinium ambiguum*)	Was a garden plant - into fields in many places. Tall blue flowers.	Causes cattle deaths frequently in USA.
Wood Anemone (*Anemone nemorosa*)	Woodland in spring. Small white flowers.	Poison: ranunculin.
Traveller's Joy (*Clematis vitalba*)	In S. Britain, hedgerow climber, creamy flowers in July/August.	Same as related Anemone.
Lesser Celandine (*Ranunculus ficaria*)	Small yellow flowers on banks & hedgerows in early spring.	Ranunculin, very irritating to the skin.
Creeping Buttercup (*Ranunculus repens*)	V. common. Like buttercup but with runners. Bad weed in gardens.	All buttercups contain ranunculin, particularly when flowering. Not poisonous in hay. *Aconite* is the really dangerous one of this family.

Plant	Description	Notes
Ivy (*Hedera helix*)	Climbing plant with glossy evergreen leaves.	Berries more poisonous than leaves Sheep can eat leaves in moderation.
Hemlock (*Conium maculatum*)	Umbelliferous. V. similar to some harmless plants. Purple blotches on stem, unpleasant smell when bruised.	Very dangerous.
Deadly Nightshade (*Atropa belladonna*)	Rather rare. Large purple flowers, June-August.	All parts poisonous.
Henbane (*Hyoscyamus niger*)	Grows in seaside places, large yellow and purple flowers June-August. White hairs on stems. Unpleasant smell when bruised.	Introduced by the Romans as a narcotic.
Woody Nightshade (*Solanum dulcamara*)	Common hedgerow weed. Purple flowers June-September.	
Foxglove (*Digitalis purpurea*)	Tall plant with pinkisk bell-shaped flowers.	Digitalis is a cumulative poison. Not less active in hay.
Honeysuckle (*Lonicera periclymenum*)	Climbing shrub with pink and cream scented flowers.	Can produce severe diarrhoea.
Lords and Ladies (*Arum maculatum*)	The Arum lily, black rod in a green sheath and scarlet berries	Berries the most poisonous part.
Ragwort (*Senecio jacobaea*)	Tall yellow raggy flowers in old pastures.	Still poisonous in hay.

SOME USEFUL PLANTS

Plant	(Human) Food	Drink	Other Uses	Medicine	Animals
ACORN	Boil twice, dry and grind to make flour, Add to wholemeal flour	Roast and grind for coffee substitute	Dyes, plus alum — use acorns, twigs and bark. Oak bark used in tanning leather	Nervous complaints Oak bark — gargle for sore throat	Acorns for pig food 8% protein 37% fat
CRAB APPLES	Boil to make jelly, apple pie	'lambswool' — winter drink	bark gives tan dye fruits give pink with alum	taken for stomach ailments	pig feed in small quantities
ASH	Seeds (keys) boiled twice and pickled in spiced vinegar		Tough elastic timber	Bark — laxative and purgative	
LADIES BEDSTRAW yellow flower	Used to curdle milk for cheesemaking		Red dye plus alum used to stuff mattresses —fragrant when dry	Popular remedy for bladder problems	
BEECH	Nuts — dry and grind for bread, young leaves in salad	Beech leaf noyan potent liqueur	Can be pressed for oil		22% protein 42% fat food for pigs
BILBERRY	Raw with cream and sugar, make jellies, pies and jam	Syrup	Purple dye	Fruit for dysentry Leaves make tea for diabetics	
BLACKBERRY	Good fruit for freezing bottling, jam, pies, jelly	Bramble wine — like port	Good dye plus alum Briars for basket making	Berries for anaemia Root cures diarrhoea	Leaves laxative
BURDOCK	Young leaf stems used in salads	Roots as coffee substitute. Dandelion and burdock drink		Decoction for skin diseases	

Plant	Food use	Food/drink use	Other uses	Medicinal	Livestock
CHICKWEED	Raw in salads or cooked liked spinach				Stops scours in pigs. Good for rabbits and hens
COMFREY (not found wild in USA)	Leaves cooked vegetable	Dried leaves make tea	Garden manure – potash. Comfrey flour and ointment sold commercially	Grated root used as plaster to set bones. Leaves applied to wounds, cold compresses for eyes	Very good food for all stock. Can form up to 1/3 of pig ration
DANDELION	Young leaves in salads	Roots roasted and ground for coffee substitute		Good tonic and diuretic	Good for pigs and rabbits
ELDER	Flowers used in pancakes, Jam, chutney	Elder flower wine Elderberry wine and cordial	Berries give purple dye with alum. Branches keep away flies	Cordial good for coughs. Wine good for sciatica	
GARLIC	Leaves in salads Bulbs in stews		Powdered, an insect repellant. Keeps mice away. Keeps birds off garden.		Treatment for worms
GOOSE GRASS	Boil as spinach. Young shoots available when little else grows	Seeds good coffee substitute	Leaves used to strain milk. Seeds used as pinheads for decoration		
HAWTHORN	Berries made into marmalade with orange peel. Also jelly	Blossom makes liqueur with brandy and sugar (hawthorn nectar) berries simmered with sprig of mint to make tea	Good quick hedge. Fuel for ovens	Heart tonic. Infusion of flowers for sore throats	

Plant	(Human) Food	Drink	Other Uses	Medicine	Animals
HAZEL	Nuts dried for winter		Hurdles, barrel hoops love philtres!	Improves heart and prevents hardening of arteries	Young shoots for rabbits
HEATHER		Tea from dried flower heads	Thatching, baskets, brooms, orange dye, fire lighting	Tea good for frayed nerves and jaded spirits	Food for sheep, goats and rabbits in Spring
HORSE-RADISH	Roots peeled and grated for sauce			Good for kidneys makes antiseptic poultice	
LICHENS	Make jelly		Some used for dyeing		
MINT	Wild mints — useful for sauce and flavouring	Mint tea from dried leaves	Prevents milk from curdling. Added also to bathwater	Warming drink Helps digestion and alleviates rheumatism	
MUSHROOM	On its own, in omelette etc. high protein food				
NETTLE	Boiled as greens in Spring. Juice used to curdle milk	Nettle beer	Fibre for yarn, oil for lamps, garden fertiliser and insecticide, green dye from leaves, yellow from roots	Rheumatism, bronchitis	Cut and made into nutritious hay for ruminants

Plant	Culinary use	Drinks / preserves	Other uses	Medicinal	Animal feed / misc
FIELD POPPY	Has been cultivated for oil in seeds, substitute for olive oil. Seeds used to garnish bread				Leaves for treating after birth
WILD RASPBERRY	Berries good raw, or jams and jellies	Raspberry leaf tea, raspberry wine		Good for childbirth problems	
ROSE	Hips in puddings and as syrup. Petals in salad and jelly. Crystallised rose petals	Rose hip syrup, rose wine, rose vinegar	Basketry for briars, petals for perfume	Flowers used to make infusion which fortifies heart and brain, good for stomach disorders	
SORREL	Curdles milk used in green salads or in stews — early Spring			Treat jaundice and liver complaints. Poultice for boils and sores	
SOW THISTLE & other thistles	Young leaves (sow thistle) used as salad or vegetable		Golden heads of some species used in decoration for for craft work. Spiky leaves to keep slugs off garden plants		Young sow thistles good for pigs, rabbits, goats and hens
WATERCRESS	Salad or vegetable			Good for anaemia antiseptic, vermitage rheumatic pains	Good for rabbits and hens

Berries are probably the most common wild food, brambles, elderberries, sloes, bilberries, rowan berries. After these, perhaps mushrooms.

Richard Mabey in his book *Food for Free* says there are at least 320 wild food products in Britain alone, excluding fish, birds and mammals. 130 of these he says are common. He speaks of "powerful and surprising flavours in store for those who venture to try them."

It seems that the wild versions of foods like cabbage are stronger flavoured than the cultivated sorts; but the foods we have grown for flavour such as onions and garlic are stronger in their cultivated form.

Another useful book recently reprinted by Prism Press, is the *Wild Foods of Great Britain* by L. Cameron. His suggestions are wide ranging, although he says they have all been traditional somewhere.

"Thus the ordinary garden snail (*Helix aspersa*) was collected all over the South Cotswold country and sent into Bristol by the ton to be sold to the working classes of that city, who esteemed it not only a delicacy, but a necessary article of food . . . their collection formed a regular occupation for the women and children of the district."

In this book are listed many things, like birds' eggs, which are no longer acceptable to most of us as food. One exception could be the eggs of the herring gull, which is reaching plague numbers in some coastal areas. To start a fashion for gulls eggs might be a good thing; I am told they are quite palatable. But of course many birds and their eggs are now protected by law.

There are other things besides food. In the uplands and wild places, the landscape is kept tame by sheep. On clumps of heather, stones and fences the sheep catch their fleece as they pass, and lose tufts of wool, especially in Spring when it is loosening ready for shearing. These bits of wool are the best material with which to learn how to spin. Hand spinning takes so much time that it is not an economic process in the sense that you could make a living from it; but it makes a very good hobby. A whole fleece would be too much for a beginner. The bits of discarded wool have no value and they are there for the collecting. Some may be dirty, but they can be washed. Vegetable dyes can be used on the wool and from these scraps a hat or a scarf or some gloves can eventually be made.

Most old country people go for walks to collect twigs for lighting fires. After a high wind, you can pick up small twigs along the lanes, with no need to trespass and you will be tidying up at the same time. There used to be an ancient peasant right to dead boughs for firewood, but if the trees are not your own, it will be better to confine yourself to fallen twigs. Longer ones will come in handy for supporting pea and bean plants.

Rushes used to be a useful yield from otherwise useless boggy land. They were used for lighting in cottages — rush-lights, which were rushes peeled and dipped in mutton fat. They can still be gathered for this if you like soft lights, or for basket making; baskets can be made from many wild plants. Soft baskets are plaited from rushes, reeds, moor-grass and so on. Hats used to be made of the same materials when straw bonnet making was a village industry. This was one of William Cobbett's areas of concern; the grasses for making many of the hats in his day were imported from Italy and he was annoyed with the cottagers for buying imported material instead of gathering wild grasses. Cobbett experimented with cutting and drying various species of grass to find out which was the best; see the account in his "Cottage Economy." Mats, chair seats and a variety of objects can be made from rushes and tough grasses.

If you want bigger things, hard baskets are made out of the pliable stems of common plants such as rose, ivy, bramble, hazel and so on. Prickly ones have the thorns removed first — rub down the stems with rags. Willow is perhaps the most popular material for baskets, such as the osier which is sometimes cultivated for the purpose. Osiers for basket making can be bought in some areas.

Bramble is a good material to start with because it is a common plant and if you take the canes in the autumn they are ready for use straight away. Some kinds of material need fading, ageing in the weather for a month or two, before they are ready for use.

Vegetable dyes, mentioned in connection with wool, can also be used to colour rushes and grasses. At all seasons of the year there are natural dyestuffs available; see the table for dyeing to get an idea of the scope. Vegetable dyes are softer than chemical ones; they fade in sunlight but they are attractive and natural. You cannot dye synthetic materials with natural dyes.

Some dyes need a mordant — a ground for the colour to stick to. The mordant is usually a mineral such as iron or chrome. In most cases the material is washed, mordanted and then dyed. The dyestuff is boiled up or left to steep for a long period and then the material to be dyed is heated up in the coloured water. Boiling is not good for wool, but the secret is to change the temperature gradually. Take it slowly up to boiling point and bring it slowly down again. Remember to keep dyes and mordants away from children and cooking pots, as they are often poisonous. Use about the same amounts of dyestuff and wool, and don't actually boil — it spoils the wool and makes some dyes dull.

The medical use of some of our herbs has been hair raising in the past and I would certainly not recommend any of the poisonous plants for use as drugs. On the other hand, it is worth finding out about the many harmless plants which can be used as simple home remedies (comfrey for example, has already been mentioned). There should be no harm in doctoring ourselves for common complaints, although we now seem afraid to do so. I would suggest J de B Levy's "Herbal Handbook for Everyone" as a good guide to the finding of plants, their preparation and the complaints they can be used to alleviate. There is a great deal of ancient wisdom in the subject and the same applies to the treatment of animals. But beware of the mediaeval herbalists and their doctrine of signatures! They thought that if a plant looked like the human lungs, that was what it should be used to treat — and so on; and they have their following still in some of the poorer herbal books now coming onto the market.

15 using milk

In this chapter I will be referring to milk from either cow, goat or sheep. In general they can be treated alike, but there are a few differences to be considered.

The Economics of Dairying

Since it can take up to three gallons of milk to produce a pound of butter, the economics have to be worked out fairly carefully; it is not so good a use of milk on the face of it as cheese, which converts at the rate of about a gallon of milk to a pound of cheese. The quantity needed for butter varies according to the fat percentage of the milk, which varies from cow to cow, most obviously according to breed.

In an old leaflet I came across the following advice;
For one pound of butter, 24 to 32 lbs of milk will be needed from Shorthorns, Lincoln Reds, Welsh Blacks, Red Polls and Ayrshires.

20 to 27 lbs of milk is the average for South Devons, Kerries and Dexters.

16 to 24 lbs milk for Jerseys and Guernseys. And of course the fat of this milk is easy to remove.

(as a rough guide, 10 lbs of milk equals one gallon).

But this is not the whole story. The skimmed milk which is left after the cream is taken off is very valuable; rather less so is the separated milk, where the cream is taken off by machine, but then the yield of cream will be greater.

Much can be done with skimmed milk. We have found

207

Jersey milk on its own to be rather too rich for our taste, used in the house; so a light skimming for butter could just make Jersey milk fit for use as whole milk.

Most of our ancestors ate only skimmed milk cheese and although it contains less fat than ordinary cheese, which tends to make it dry and hard, it is an excellent way of getting the value from the milk and would make a good cooking cheese.

The cost of milk production need not be too high because a spring calving cow can get enough goodness from grass for summer milk production; we need not depend on expensive concentrates. The expense of buying the cow in the first place is large, unavoidably, and of course there is always the risk of losing her, as with any animal. But the cost of cows is related to their production and a beefy cow will give you a good saleable calf every year and cut out all your dairy bills entirely; she will live for a good many years — one hopes — and as a long term investment the enterprise does make sense.

Composition of Milk

%	Total Solids	Fat	Casein	Other Nit. subs.	Lactose	Ash
Cow	12.46	3.71	2.63	0.66	4.70	0.76
Sheep	19.18	6.86	2.63	1.55	4.91	0.87
Goat	14.29	4.78	3.20	1.09	4.46	0.76

The most obvious difference is the colour. Jersey milk is yellow; most cows' milk has enough colour to give a pale golden butter in summer, although some breeds produce more carotene than others and this gives the milk its colour. Goats' and sheeps' milk is dead white and although white cheese is acceptable, white butter looks odd. This can be overcome by using the standard dairy vegetable colouring, which is annatto. It is annatto which gives red cheese its colour — the only difference between red and white cheese. Home cheese makers used to use marigold petals and carrot juice to colour their dairy products.

Another difference is in the size of the fat globules. Milk is an emulsion, a liquid with small particles of solid suspension — the fat. In goats' milk the particles of fat are small and this makes them more digestible. It also makes for difficulty in separating the fat from the rest. Some breeds of cow tend to give milk with small fat globules, for example the Ayrshire; Channel Island breeds give milk with large particles of fat which rise quickly to the surface. Milk with small fat globules is good for cheese making because the aim of the process is to trap the fat in the curd rather than let it escape in the whey. And of course, the creamy milk with large fat globules is easier to make into butter.

Sheep's milk is very rich, which makes it good for butter making; care must be taken when making cheese with rich milk. Commercial dairies actually put whey through a cream separator to recover the fat they lose in the process. This is made into "whey butter" and sold for cooking purposes and cosmetics.

Dairy Equipment

This can be as simple or as sophisticated as you like. Until recently it was difficult to get hold of equipment for making cheese and butter on a small scale, but suppliers have begun to stock these things again — see the back of the book for details. Second-hand equipment can often be bought at farm sales.

Cheese and butter have been made for thousands of years and before thermometers were invented the dairymaid's elbow was the instrument used to measure the temperature. But the quality of the milk products of the past must have been variable in the extreme; not all those picturesque old dairies were hygienic. We have very sensitive palates compared with our ancestors, which is why a lot of backyarders are disappointed with their first efforts at dairy produce.

We are used to clean, if uninteresting, dairy flavours and we want to be able to follow cheese recipes exactly so that if we hit on a good one, it will be reproduceable! So the dairy thermometer will be a vital piece of equipment, and the rest of the things will be of good quality, stainless steel and seamless and thus easy to sterilise. Milk is difficult to get off equipment without taking trouble, and if left on will be a perfect breeding ground for bacteria, which will foul up the next batch of milk. It always pays to get good dairy equipment.

Strainer. Most people strain the milk; if nothing else, it is a test of cleanliness, because sediment in the strainer indicates bacteria in the milk. The bugs are washed through the filter pad, leaving the debris behind. Be sure to strain if you are a learner, or teaching someone else.

One reads of goosegrass being used to strain milk, and mares' tails weed for scouring the pails; these are nice bits of folklore, but not really practical. Traditional sieves (or "siles" in the north of Britain) are rather like a vegetable colander, with two handles, but instead of being covered in holes they have two metal perforated discs in the bottom, between which a filter pad is placed. Before these came into use, milk was sieved through butter muslin stretched over a pail.

Coolers The cooler is important, because as soon as it leaves the cow the milk should be cooled. This slows down dramatically the rate of growth of the bacteria that succeeded in getting in — and there always are some. Small quantities of milk can be cooled by standing the container in cold water, but this is only effective in summer if it is running water. The cheapest cooler we ever had was a mountain stream; some old dairies were built over streams. This is a much better idea than the regrettable one of building your loo over the local rivulet; but I think that some newcomers, finding a small hut over a stream, have mistaken its purpose. Ten to one it was a dairy.

Commercial farmers are almost all changed over to refrigerated cooling in bulk milk vats, so there should be surplus coolers on the market now. Basically these old coolers were of two types, the in-churn and the surface cooler.

The in-churn cooler consists of a head to fit over a milk churn, with metal pipes running down into the milk and up again. The head is connected to a water hose and the force of water going through the metal pipes cools the milk and turns the pipes, which gently agitates the milk. Then the water runs down over the outside of the churn, and is wasted unless you stand the churn on a grid over a tank. In this case the water can be used for washing down and so on, but if it is to be animal drinking water the churn must be very clean on the outside.

The surface cooler is usually on a stand high enough for a ten gallon milk churn to stand on the floor underneath it. The milk is poured into a tray on the top; it trickles down the corrugations and is cooled by the flow of water inside the corrugated case. The rate of flow of the milk can be adjusted,

so you can get it fairly cold if you go slowly enough. The
water hose is connected to the bottom of the corrugated
case and the warmed water comes out at the top and can be
piped into a tank for further use. Mains water is sometimes
rather warm in Summer; the surface cooler is the more
efficient in this case, but it cannot cool lower than the temperature of the water. Brine solution will get the temperature
lower.

Milking Buckets

Wooden buckets look picturesque, but don't use them;
they are impossible to sterilise properly. Hooded pails for
milking are excellent if you can get them, they do cut down
the amount of debris falling into the milk during hand milking.
All surfaces coming into contact with milk should be seamless; stainless steel is best, but expensive. Never use a pitted
metal bucket, and watch all metal equipment for deposits
of Milkstone (see cleaning).

Strip Cup This is something you can make for yourself.
They are simple enough, a metal cup with a handle and a
black disc, which has a small piece cut out of it so that milk
on the disc can run down into the cup. The idea is to take a
squirt of milk out of each quarter onto the disc, and look at
it for any abnormalities — clots particularly or blood, which
will indicate something wrong with the udder.

This is actually obligatory for commercial herds, being
written into most countries' dairy regulations. It can be useful for backyarders to keep a check on the health of the
udder. In any case, it is a good idea to discard the first squirt
of milk from each quarter because it has more bacteria in it
than the rest, bugs that have crept into the teat since the last
milking.

Milk Churns (a misnomer — these are not used for churning) They are 10 gallon metal cans which were once used to
send milk to the creameries, before it was collected by bulk
tankers. Now they are going out of use, and should be available for backyarders — see your local creamery manager.
They are the best utensils to keep milk in, and essential with
an in-churn cooler. If you buy one, make sure that the inside
is not pitted or rusty.

Cream Separating The commercial electric separator has
many little steel discs and takes a long time to wash up; there

is a smaller one available but the principle is the same. Goats' milk producers will probably need a proper separator, either worked by hand or electricity, because of the difficulty of removing the cream by skimming.

Separators work by centrifugal force; the cream is lighter than the milk and is flung to the outside of the discs, where it is collected into a cream spout. It comes through one outlet and the separated milk through another. The cream screw adjusts the rate of flow and thus the thickness of the cream.

The age old way of getting the cream off the milk was to set it in wide shallow pans in a cool place — the cooler the quicker. A perforated disc is used to skim off the cream when it has risen. You may consider this wasteful in that about ½ a percent of cream is left in the milk, but with Jersey or other rich milk you can skim off some of the cream and then use the skimmed milk like ordinary milk, which means the cream is a bonus. In some places, milk was set in shallow lead creamers and after the cream had risen, a plug was pulled and the milk ran off from the bottom, leaving the cream stranded. We know now that it is dangerous to use lead equipment.

Butter Churns Old wooden churns at farm sales look attractive, but sometimes they are sold for more than they are worth. We have one such museum piece which turned out not to have a paddle in it, and it was once a paddle churn. You would need a great deal of cream for one of those old churns. Either the cream is knocked about by falling from one end of the churn to the other in an end-over-end churn; or else it is knocked about by a paddle in the middle. The shock is needed to encourage the coalescing of the butter globules.

For small quantities of butter an electric food mixer can be used. There is also the Blow churn, a glass jar with a wooden paddle turned by a handle at the top. Since it is glass you can see the whole process; these churns are available in several sizes.

Butter Workers are not essential unless you are intending to make a large amount of butter. The worker is a sort of wooden board, with a slope on it, and a wooden roller. It is used to work the butter, expressing the moisture and pressing the grains of butter into a solid without breaking them — the essence of good butter making. Rough handling of butter when working causes the grains to smear out and the butter is then greasy; so the job must be done properly. Small quantities can be worked with the Scotch hands, which

will be needed in any case to form the worked butter into pats. These are usually made of beechwood, bats with handles and grooved surfaces.

Cheese Equipment I learned to make cheese in a 50 gallon vat, with a metal jacket round it; steam was injected in to the jacket to raise the temperature of the milk. This kind of cheese vat would be too big for most backyarders, but the traditional cheese recipes were intended for this type of equipment, used in farm dairies about fifty years ago. For small amounts of cheese the kitchen fire and a "cheese kettle" had to do the job. For these we can substitute the kitchen stove, and a large saucepan or preserving pan. An asbestos mat will help to tone down the heat a little.

Hard cheese needs pressing, and the containers in which it is pressed are called moulds. They can be bought, or you can improvise with recycled tins eg large coffee tins, with holes knocked into them for drainage. A wooden board can be cut to fit the top as a "follower". It should be slightly smaller than the mould, taking the pressure down onto the cheese when the curd sinks in the mould. Screw down cheese presses exert tremendous pressure; they are now museum pieces and are bought by collectors who will never make cheese. A heavy weight will do instead, or a car jack, braced against something firm.

For soft cheese making, a ladle with a sharp cutting edge is needed, and some small metal hoops — Coulommier hoops. Imitations of these can be made by cutting down tin cans and bending over the edges to make them safe. Straw mats are also needed; the cheese stand on them to drain.

213

In general, it seems best to start with the minimum of equipment and see how you get on with making dairy products. The best of home made cheese and butter is superb, but the worst is horrid and if you are not inclined that way, it may be better to forget about an involved process such as hard cheese, which after all does take time and skill. Milk can still be a blessing to the family even if you only make say soft cheese and yogurt, and freeze the surplus, or feed it to the pigs.

Treatment of Milk

Milk is perfect food for bacteria and life in the dairy is a constant battle between them and us. Immediately after milking, strain the milk and cool it as quickly as possible. If there is no cold running stream in which to dump the can, buy a cooler; its worth it. A small amount such as the milk of one goat can be cooled by immersing the bucket in a large container of water, if the water is very cold; chunks of ice or salt in the water will help in Summer.

Beware of taints. A cool place for storing the milk is essential and it should always be covered. Never store it near strong smelling things like onions because it absorbs smells very easily.

Washing Equipment

All dairy equipment must of course be very clean to avoid a build up of the dreaded bacteria, which can easily live on unclean surfaces and spoil your next milk. This is the drill, and it pays;
1. Immediately after use, rinse utensils in cold water. Hot water makes the protein stick on.
2. Wash thoroughly in warm water, with something in it to dissolve the fat; washing soda is cheapest and best for this. If you use cold water at this stage, traces of fat will be left.
3. Sterilise. Do this by immersing in boiling water for two minutes. Big things like butter churns have boiling water poured over them. Don't make the mistake of boiling the thermometer!

Dairy farmers now use chemicals such as hypochlorite or iodophor to sterilise their equipment. Boiling water or steam is better, and just as easy for backyarders. If you do use chemicals, never mix them. And do not use them for cheese and butter making equipment as they may leave traces which could spoil the product.

"Milk Stone" is a hard yellow deposit on metal, a scale which is worse in hard water areas; it builds up on dairy utensils which are less than perfectly cleaned and sterilised, but most of us get it at some time. Acids are used to remove it commercially, but if you are reluctant to keep concentrated acid about, the Iodophor type of dairy cleanser which is acid-based, will gradually clear the milk stone; in fact this type of cleaner will show you whether the milk stone is present — the yellow colour of the chemical will persist after washing.

Milk Products

Yogurt. I begin with this because it is one of the easier milk products to make and a quick way of dealing with milk which is surplus to liquid requirements. This is the best form in which to try to sell goats' milk to those in search of health foods, because it will keep longer than fresh milk. The healthy and long lived inhabitants of the Balkans ate a great deal of yogurt.

Yogurt consists of a culture of acid producing bacteria which you have induced to make a home in your milk; they attack the lactose or milk sugar and turn it into lactic acid. This preserves the milk for longer and it also benefits our digestive system, which can be short of acid. When we have taken antibiotics, for example, the beneficial organisms of our intestines are killed off with the bad ones, and yogurt helps to replace the good ones again.

To make yogurt, heat the milk gently — in a double pan if you like — to just under boiling point and then allow it to cool to blood heat. (The Holder method of pasteurisation, as used in industry, is 145^OF for 30 minutes, and the HTST (high temperature short time) method takes the milk up to 161^OF for a brief time. But if you object to heat treating the milk, this can be left out. It merely kills the natural bacteria present in the milk and allows more room for the yogurt bugs. Unless you fear disease in the milk, there is no real need for pasteurisation.

Having got the milk to blood heat, the next job is to introduce the yogurt organism — the one we used to use was *Lactobacillus bulgaricus*. It produces more acid than the cheese starters. The easiest way to do this is to tip in live yogurt of somebody elses make; the rate of about a tablespoon of yogurt to a pint of milk is about right for pasteurised milk, but if the milk is raw, there will be more competition and more yogurt may be needed. I have even done this with ordinary sweetened fruit yogurt from supermarkets, but the plain health food sort will be best. If you are a perfectionist, you could obtain some culture of the organism from dairy suppliers and propagate your own.

The next stage after mixing it up is to let it develop for about 24 hours, keeping it warm. Bacteria multiply faster in warmth up to the point where it gets too hot for them — somewhere over 105°F. This is the basis of much of our dairying.

Keeping yogurt warm is easy with an all night stove or radiator; otherwise, use a big vacuum flask with a wide mouth; or perhaps a haybox, which sounds a good idea — I've never tried it. A wooden box filled tightly with hay, is insulated and acts like a vacuum flask.

Your family may have acquired a taste for bought yogurt, in which case add honey, nuts or fruit to bring yours up to their expectations. Plain yogurt is an acquired taste, but you can get hooked on it.

BUTTER MAKING

Butter can be made as soon as the cream is skimmed or separated; but if you wait until the bacteria have developed a little — this is called "ripening" — the butter will be easier to make and it will have a more interesting flavour.

Artificial ripening

Those who make butter in large quantities ripen the cream artificially, but backyarders can let it happen naturally. There were two ways in which we ripened cream, useful to know if you are in a hurry!

1. Add buttermilk from a previous churning which will already have a degree of acidity. It must of course be from

good butter to give the right flavour! Half a pint of this to a gallon of cream is about right.

2. Pasteurise by heating the cream to 150°F and then cool and add starter. Butter starter is a culture of bacteria most likely to produce a good butter flavour — not the same as the cheese organisms. They can be bought as culture for propagation, or begged from a butter-making creamery.

Natural Ripening

This merely means allowing the milk bacteria to develop and enhance the flavour. The cream is kept two days in summer and up to four days in winter, covered with a muslin cloth, in a clean place and at a temperature of 56°F to 66°F (For the scientists, when the cream is ripe the lactic acid percentage should be about 0.5%).

Getting The Cream Ready For Churning

When you are ready to churn, the cream must be at the right temperature and consistency. To warm it up, stand the cream in its container in a bucket of hot water; use cold water or brine if the object is to cool it. The churning temperature was one of the more critical things in life when I was learning things; it varied every day according to the dairy temperature. (See the table). Butter making dairies are not heated!

Dairy Air Temperature deg. F.	Churning Temperature deg. F.
62	53
60	54
58	55
56	56
54	57
52	58
50	59
48	60

You will see that as the weather gets colder, the cream is slightly warmer when churned.

If the consistency of the cream is right, it will run freely off the Scotch hands (the wooden bats you use to handle the butter with). During ripening, the butter is stirred once or twice a day, and it should be all the same consistency when churned. If it is too thick, add a little water at the right temperature.

The exact fat percentage of the cream does not matter too much, but thinner cream seems to churn more quickly in summer. In Wales, I found that traditionally whole milk, slightly soured, was churned instead of cream. This is an interesting variation, hard work (unless they got a dog or water power to turn the churn) but producing lovely rich buttermilk. If your Welsh friends go into raptures about buttermilk, this is why. It is really skimmed milk and buttermilk combined and is drunk and used in cookery.

Traditionally, the cream ready for the churn is strained through muslin into the churn, which should not be filled more than half full. A wooden churn is rinsed out first with water, so that the cream does not stick.

Churning

Whatever kind of churn you use, the cream is bashed about, slowly at first. With the end over end churn you work up gradually to about 55 revolutions per minute. The process will take about fifteen minutes or longer with a wooden churn — up to 40 minutes of hard work! Sometimes nothing happens; you have "sleepy cream"; a tip is to add a little water at about 70°F in this case. When it does work you can hear the sound change as phase inversion occurs; the fat becomes visible in globules and the liquid is thinner and bluish. During the process, ventilate from time to time with the old churns — there is a little button in the lid of a wooden churn, which you press to expel the rather smelly gas that comes off ripened cream. In the lid there is also a sight glass through which you can peer. Using a glass churn is simpler because you can see what happens and there are holes in the lid for automatic ventilation.

As soon as the swish of the cream churning changes to a thud, the butter has broken — stop churning. The grains of butter should be small like mustard seeds; if you go on after this, "footballs" will result, and the butter will not be so good. But this is the professional approach — don't worry if you get just one huge football, you have still succeeded in making your own butter.

Then the buttermilk is drawn off through a fine sieve; with a big churn it comes out through the bung in the bottom.

Washing The Butter

Water is poured into the churn and the butter is sloshed around a little; this is done twice. In hot weather, the water should be colder than the butter to firm up the grains a little. When it is very cold, slightly warmer water will make the butter softer for working.

Some people salt at this stage, using brine. 1 lb of salt is dissolved in 1 gallon of water and this is poured over the butter, which is then left standing in the brine for about 20 minutes.

Dry salting is more usual; add about ¼ oz salt per lb of butter before working, or more to taste.

With the bigger churns, there is a perforated concave scoop for fishing out the butter grains from the water, which prevents them from sticking to the side. I have yet to find an elegant way of getting butter out of a small glass churn.

Working Butter

The object is to squeeze out surplus moisture without injuring the butter; clumsy handling will result in a greasy product. Press with the Scotch hands, do not smear. The legal maximum water content for butter is 16%. The water squeezed out should be quite clear; if it is milky, the butter has not been washed properly and will not keep so well.

The buttermilk, which you have carefully kept, is a traditional drink for the hayfield, or poured over oatcakes to to make a Welsh delicacy; it was also used in Wales for washing slate floors, because it imparted a particular sheen!

Potting Butter

This is a method of storing for the winter — apart from freezing, and of course butter freezes very well.

For potted butter it is usual to add 1 oz of salt per pound and the butter is worked very thoroughly to get as much moisture out of it as possible. It is packed down tight in a glazed earthenware "crock". The last layer is a thick layer of salt, or brine. Then a piece of greaseproof paper is tied tightly over the top. Keep it in a cool dry place and it should store well for several months. When it is used, soak it for a while in water at $65^{\circ}F$, and rework it, sprinkling with cold water to freshen it.

CHEESE MAKING

There are thousands of cheese varieties, but the basic principles are the same and they are simple. The milk is warmed and made to form a curd, usually by the addition of rennet, which is an enzyme found in a calf's stomach. Then the curd is cut into pieces and cooked by stirring it in the whey (or taken straight out to drain in the case of soft cheese). The whey is run off, and the pieces of curd stick together to make a solid, which is further drained and cut into pieces again before being packed in moulds and pressed. Soft cheese is just curd, with little cooking and no pressure. It is only slightly acid and will therefore only keep for a short time. As cheese gets harder, the acidity and the cooking temperature are higher. Cheddar is our hardest cheese.

Starter, the culture of acid-producing bacteria added to give the cheese the right acid development, texture and flavour is optional; but you may have to get used to a different and variable sort of cheese. Commercial dairies invariably pasteurise their milk and replace the natural milk bacteria with starter because they must have a uniform product. But our ancestors managed without starter, and so can we. You could try using yogurt as starter — it will be similar to a cheese culture although not identical.

Rennet will be needed except for the very simple lactic cheeses which are simply made from milk left to go sour and form its own curd. But rennet is obtainable and will keep, so this is not really a problem. There are herbs which have been used by country people instead of rennet to coagulate milk; nettle juice, ladies bedstraw and sorrel spring to mind.

The standard cheeses such as Cheddar and Cheshire used to be local cheeses — these particular ones became famous because they travel well. The purists say however that they

are only at their best when made on home ground; that the salt grazings of Cheshire give that cheese its tang, and the limestone pastures of the Ure valley are necessary for the proper production of Wensleydale cheese. If you subscribe to this theory, which incidentally seems to be gaining ground in scientific circles, you will want to make the cheese which belongs to your area. A drawback to this is that they developed as farm cheeses, made from the milk of a herd of cows or even several herds — Cheddars were huge, and a co-operative effort. A bigger cheese was made than the backyarder can manage and the size helped it to mature properly. Small cheeses tend to dry out when stored. So you cannot really make a small Cheddar, although one pound tourists cheeses are made by creameries these days, so there is no law against it. But I feel that hard cheese making takes so long, I would not like to go through the whole process with small quantities, except for Wensleydale, traditionally a rather small cheese.

These are the reasons why you see recipes for "smallholders cheese". Do not be discouraged from trying hard cheese if you fancy it, but also, don't worry if it is not quite like the produce of the factory.

Smallholders Cheese

"Five gallons of fresh sweet milk are required to make one cheese of standard size, and the circular cheese mould

221

should be 10 inches in diameter, 4 inches deep and perforated. (Holes in a cake tin?) It should be provided with a
closely fitting wooden disc or "follower". The milk is first
raised to a temperature of 92°F to 95°F, and rennet in the
proportion of 1 dram to each two gallons of milk is added.
The rennet is diluted with water about six times its volume
and mixed in with the milk.   The tub is then covered with a
lid and left for 30 to 35 minutes. (Keep stirring gently on the
top until you feel it begin to thicken — if rennetted milk is
left too soon, the cream will rise to the top and later it will
be lost in the whey.)

Cutting the Curd

The curd is ready to cut when it feels firm and springy and
splits with a clean fracture in front of the finger. A large
knife, long enough to reach the bottom of the tub, is taken
and the curd is carefully cut into ½ inch sections. When cutting is completed, the curd is gently broken up and stirred
gently by hand for about 10 to 15 minutes, then allowed to
settle in the whey.

Scalding the Curd

A cheese cloth is thrown over the tub and pressed down to
the curd, a quantity of whey is thus ladled off into a pail, and
the temperature of this is raised by immersing the pail in hot
water to such a degree that when it is poured back into the
tub, the contents are heated to about 98°F. This is known as
scalding the curd (in proper cheese vats, steam is injected into
the jacket to raise the temperature).

There is a formula for working out how to be accurate
with the temperature:

$$\frac{\text{Gallons in tub X degrees to be raised}}{\text{gallons removed}} + \text{temp of whey when removed}$$

= temp. to which whey should be heated.

After scalding, stir for 20 to 30 minutes, by hand or with
a wooden spoon. The curd will go tougher and brighter and
sinks rapidly. On pressing a little in the hand, the particles

222

should not break and should be a little springy. When this stage is reached, the curd is allowed to settle for 10 minutes, (pitching) and the whey is then poured off through a straining cloth; try to keep the temperature up to 98°F if possible all the time.

Moulding and Pressing

The mould is placed on a board and a cheese cloth placed inside; the curd is then taken out of the tub, broken up and filled evenly into the mould. When filling is completed, the edges of the cloth are turned over, the wooden follower is placed in position and pressure is applied at once by means of a 14 lb weight, which should be kept on for 15 minutes. The weight and follower are then removed, the edges of the cloth turned back, and the cheese turned by hand. Hoop, cloth and follower are then replaced, and the cheese weighted up to 21 lb. In 30 minutes more the cheese is again turned and 28 lb weights are applied. Thus it is left for 4 hours, when the cloths and follower are removed, the edges of the cheese are trimmed and it is left uncovered in the mould.

Salting and Ripening

The next morning salting takes place. Rub 1½ oz salt carefully all over the cheese, leaving a little extra salt on the upper surface. It is again turned and salted in the evening, and the following morning washed with 10% brine (I lb salt per gallon of water) and placed on a shelf to dry. On the third day after making, it is taken to the curing room or cellar and turned each day until ripe. It may be rubbed occasionally with a little brine to keep the skin clean. At the end of three weeks the cheese should be ready for use, but it will improve in quality if kept for six weeks."

I have included this recipe in detail because it is possible to use it to produce good cheese with the minimum of bother and no specialised equipment. This is a basic cheese recipe, but there is much more work to the hard cheeses — the curd is chopped up and weighted and so on; the smallholders cheese is comparatively simple — I hope you try it.

Certain faults can occur; if the cheese ferments and gets

223

CHEESE VARIETIES

Cheese	Starter	Rennet	Curd
CHEDDAR Originated in Somerset hundreds of years ago. The milk from whole herd often made into one big cheese. Can be white or coloured 30 lb cheeses	Rather more in winter than in summer. Average about 1½% of milk quantity Use morning and evening milk, Warm up cream separately from evenings add at 80° F milk	1½hrs after starter added. Warm milk to 84°F or a bit higher in cold climates. 1oz to 25 galls milk diluted with 3 or 4 times amount of water and stirred in for 4—5 mins. then just top stir to keep cream in until it starts to set	Ready to cut when breaks cleanly over finger Slice up with large knife or use vertical and horizontal cheese knives into ¾" cubes
CHESHIRE made in Cheshire since at least 12th Century. 30 lb cheese usual	Warm milk to 86°F and add about 2% starter Ripen for 1—1½ hrs before adding rennet	A little less rennet than for cheddar	Ready to cut in about 30 mins. as above. Cut into pieces the size of a haricot bean. Very careful cutting to avoid loss of fat
CAERPHILLY takes name from Glamorgan village Miners took this cheese to work with them. Smaller cheeses 8 lb	Milk at 70°F add 2% starter and leave 1 hr. to ripen. Take temp. up to 88—90°F	90°F add 1oz to 25 galls and stir, then leave 30 mins.	Cut into ¼" cubes and stir 10 mins.
DERBY more a creamery cheese than farmhouse	Milk at 70°F 1½% starter added then heated to 84°F and acidity raised to 0.2%	1½oz to 25 gall milk stirred then left 40 mins.	Cut into ½" pieces allow to settle then stir

SAGE DERBY interesting variation — layer of sage leaves etc, cabbage for colour

Scalding & Pitching	Milling Moulding Pressing	Storing	Remarks
Stir until curd is very firm, raising temp over about 45 mins. until it reaches about 102°F Pitch —leave to settle in whey. Drain whey off when acidity above 0.18%	'Cheddaring' is cutting mat of curd into 12" blocks piling them two high and keep turning while acidity goes up. Keep warm. Tear curd into small pieces, salt and push into moulds. Press as hard as poss 80 hours	Bandage. Turn every day for a fortnight keep at 50°F. After a month even cooler if poss. Brush every week Humidity 84% in store room	Keeps very well and for a long time. Ready in about 6 weeks
Stir and after 15 mins raise temp gradually to 88°F over about 45 mins. Pitch when curd particles retain shape when squeezed. Acidity 0.15%. Pitch for about 30 mins. with acidity 0.17%	Cut into blocks and then break them in two by hand. Turn every 20 mins. for 1½ hrs. Mill at 0.60% If curd is firm and dry. Sprinkle with salt and put into moulds. Pressed down 15 mins and remoulded. Over-night pressed 25wt	Capped and bound with calico bandages Turn every day for a fortnight	Ready in about 3 weeks
Heat 92-94°F and stir until acidity 0.14-0.17% about another 30 mins Settle 15 mins and run off whey Pile curd to sides acidity to 0.24%	¼ lb salt to 25 gall milk mixed in and packed quickly into moulds. Light press up to 2cwt for 2 hrs then reversed	Floated in brine 24 hrs then shelved and turned every day. Not bandaged	Ready 3-4 weeks Does not keep very long
Take temp up to 94-96°F over 45 mins. Pitch until 0.22%. Draw off whey quickly cut into 6" blocks and pile to sides of vat. Cut again every 20 mins.	Break into large pieces at 0.6% acidity 2% salt added and mixed in mould and press lightly cap and bandage next day	Repeat at 60°F Turn daily	Sage Derby an alternative. Layers of sage into moulds See note.

Cheese	Starter	Rennet	Curd
DOUBLE GLOUCESTER from the Vale of Berkeley should be made from milk of Gloucester cows now a rare breed	84-86°F for about 1 hour may use starter but not essential a little annalto added	Under 1oz per 25 gall stirred then left 45 mins. with milk at 86°F	Cut into 1/8" cubes stirred 15 min. acidity 0.13%
LANCASHIRE supposed to be cheese and butter in a sandwich	Heat up to 85°F add starter 1oz to 5 gall	Rennet after 40 mins. Leave for 40 mins. 1 dram/ 2½ gall	Cut to size of small bean
LEICESTER deeply coloured red	70°F starter at 2% Annatto 1oz/ 10 gall added and stirred well	Milk up to 85°F rennet ½oz/10 gall leave for 40–50 mins.	Cut into pieces size of a grain of of wheat, acidity to 0.12% stir
WENSLEYDALE softer, French in origin. Originally produced from sheep's milk by Cistercian monks Might be useful for sheep's milk now	70°F starter 1%	Milk up to 84°F less rennet than for others. Keep covered before cutting. Leave 40 mins.	Cut into ½" cubes stir and raise temp to 90°F

The harder cheeses are rather more difficult to do well with high fat milk such as sheep's or Jersey milk. Try a softer one with very gentle handling of the curd.

Scalding & Pitching	Milling Moulding Pressing	Storing	Remarks
Heat gradually to 94-98ºF over about an hour stir till curd feels notty 0.18% acidity Curd pushed to end of vat and whey run off. Cut into 6" blocks. Turn every 20 mins. Can put on a rack and weigh them to express whey	2 hrs later, 1-1¼ pounds on hot iron broken into small pieces and salted moulded at 76ºF Press 2 days 10cwt from 20cwt turn night and morning with clean cloth	Ripen at 50-56ºF and turn daily	4-6 months for proper maturity
Soon after cutting curd allowed to settle and whey drained off slowly	Curd cut into blocks and weighted to get out whey. Left in blocks till next day in warm room	Next day process repeated and curds are mixed — 2 days making — and moulded after salting at rate of 1oz/3½lb curd. Pressed 5-10cwt next day for 2 days Bandage on 3rd day and press again for a day	Can be 3 days curd, 1/3 of each in cheese
Increase temp to 93ºF in 40 mins then allow curd to settle put on racks and weights to consolidate 0.18% draw off whey. Cut curd into 6" blocks and pile to sides of vat	Cut and repile every 20 mins to 0.45%. Break into small pieces Add salt ¼lb/50 galls stir in very evenly. Press gently at first then increase to 10cwt	Press next day again, then cap bandage fine cloth Turn every day	
Stir 15 mins until pitching 0.16%. Draw off whey gradually. Cut 3 or 4 times into smaller blocks	Milk when curd soft and flaky about 0.5% salt ½lb/50 gall Put into moulds leave 1 hour then put into cloths and press		

To make cheese seriously you need an acidometer to determine acidity. Whey is titrated against N9 NaOH using phenolphthaleine as an indicator.

out of shape during ripening, this means the wrong sort of bacteria are present; remedy, pasteurise the next batch of milk and use starter. Should it go hard and crack, this means that too much fat has been lost in the process — rough handling of the curd can cause this, or not stirring long enough after the rennet goes in.

Cambridge Cheese

There are many soft cheeses, made for rapid sale and immediate consumption. These cheeses were only sold in summer, often on a fresh cabbage leaf.

The temperature of the milk at renneting should be about 92°F, say about six quarts of milk can be used; ½ dram of rennet is diluted and added as before (no starter needed).

When the curd is well set, the whey comes to the surface. Then the curd is slowly and carefully removed with a sharp edged ladle and placed in the moulds in thin slices. The moulds are in two pieces, the bottom standing on a straw mat while the upper one has draining holes. The moulds are about 7½ inches long by 5 inches wide and 6 inches deep and used to be made of elm wood. The wooden moulds keep the heat longer than metal, but round metal hoops can be used for this type of cheese.

These cheeses are not turned at all, and are ready to eat when the wooden moulds can be removed without the cheese losing shape. Keep them on the straw mat. If the cheese is tough and leathery, the temperature has been too high or the drainage too fast; if soft, the temperature was too low.

Whey

Whey is a valuable animal feed, particularly for pigs. 1¼ gallons of whey equals 1 lb barley meal in feed value.

16 using meat and skins

Now that freezers are possible, home storage of meat is no
real problem; but not everyone has electricity, so salting and
drying can be useful alternatives. The provision of a steady
supply of meat comes under the heading of Planning, and
nowhere is it more necessary.

We can divide our meat production into two kinds; small
quantities, which can be eaten fresh and killed only when
wanted, and larger animals which must be stored in one or
more ways. The smaller ones include poultry, rabbits and
fish. Sometimes we may need to harvest a batch of these
together and either freeze or salt them, but in general they
will provide just one or two meals at a time. The planning
in this case involves not having too many ready at once.

Eating meat in season is no longer fashionable since the
freezer has made all seasons alike, but backyarders know the
particular appropriateness of all the foods in their proper
season and it is one of the treats of the job and reward for
our labour that we can revive some of the old traditions.
Michaelmas geese, for instance; they had been grazing grass
all summer and finished off on the corn field stubbles. At
Michaelmas the geese were ready to eat. Beef was once supposed to be at its best when French beans were ready;
Mutton was at its prime when the grass was most plentiful.
So — don't get carried away by the freezer and remember
to eat some of your produce when it is fresh!

Nutritionally and to the palate, meat is better fresh. Freezing is the best method of storage, because after the meat has
been cooked there is little difference in value between fresh

229

and frozen meat. Nothing is lost during storage, but when it thaws, some soluble nutrients may be lost in the water that drips out. Smoking and canning cause loss of thiamin and a deterioration in the quality of the protein. Salting and drying are not so good nutritionally; they do cause loss of nutrients; and of course too much salt in the diet can be harmful. Most meat can be desalted to an acceptable level by soaking in water before cooking.

A variety of food is best because in variety we get all the elements of the diet that we need. Meat varies widely in value — see the table.

Average composition of cooked meat	Protein, g. per 100g	Fat
Bacon	24.5	38.8
Beef (average)	18.1	17.1
Black pudding	12.9	21.9
Chicken	24.8	5.4
Ham	24.7	18.9
Kidney	16.2	2.7
Lamb	23.0	22.1
Liver (fried)	24.9	13.7
Pork	15.8	29.6
Tripe	9.4	2.5

It is only reasonable to consider the value of cooked meat, although the value of it raw is often quoted — but nobody eats it like that, except for fish in some countries. The effects of cooking are several. The muscle protein, myoglobin is changed by heat; it goes brown at temperatures over 65°C. The fibres coagulate and the meat becomes firm. Shrinkage causes loss of some of the juices. Of course weight is lost; as much as 30% of their original weight is lost when chops are grilled.

About a quarter is lost when joints are roasted, and bacon is reduced to half its weight when fried, that is if it is cured in the modern way with brine.

Cheaper cuts of meat should be cooked slowly and with moisture, to allow the collagen in the connective tissue to be converted into gelatin which is much more tender and more valuable.

Animals for meat should be killed in prime health; they should not be breeding or moulting. The exact size will not be critical for a backyarders' dinner, as it would be if the animals were being produced for sale, but economic laws will still operate and in some cases there is an economic time for killing. In the case of home produced beef it may be when the grass gives out in the autumn; ducklings over about ten or twelve weeks will eat more than the extra meat they produce will be worth.

Before slaughter, animals should be fasted, but should be allowed plenty of water and should not be alarmed or excited. This is easier when the animals are killed at home; they need have no idea of what is happening. It is more humane, and there are also practical reasons for keeping animals quiet; this is particularly important for pigs, especially for bacon.

Fasting is important because if the blood contains food substances which are partly assimilated, this may affect the meat. Nervous excitement causes a rise in temperature which means that the carcase will not bleed properly. If the animal is tired, some of the glycogen reserve in the muscles will be used up. When an animal is slaughtered, these reserves change to lactic acid and this helps to preserve the meat. Recent research suggests that a little sugar should be fed to pigs immediately before slaughter.

Slaughter of the larger animals such as pig and sheep at home is now legal in Britain, and may be necessary in an emergency. There are still plenty of older countrymen and women who know how to go about it. But in general the local abattoir will do the job — they actually have an obligation to do so — and they will give you back a dressed carcase, jointed or not as you prefer. Poultry and rabbits can be killed and dressed at home, but only for family consumption. If a surplus is sold it must be sold live.

In the other backyard books are detailed accounts of how to deal with the various kinds of animal; even with the larger ones, it may be as well to know how to do it! Briefly the drill is this: shoot, bleed from the throat immediately, sling up, slit and disembowel. Then you can take your time with the rest. The pig is rather more difficult because as soon as it is killed and bled, it is scalded and scraped to remove the bristles, and this is hard work.

Poultry and rabbits are killed by dislocation of the neck, after which they are skinned or plucked and then the insides are taken out. The knack can be learned best by watching someone do it.

As soon as possible the meat should be cooled to help the keeping quality; like milk, it is a prime breeding ground for bacteria and all possible hygienic precautions should be taken whenever meat is being handled. Small animals are easy enough to deal with and to cool, but a whole beef animal can be a nightmare without proper preparation, unless it comes home dressed, when it will at least be cool to start with. If it so happens it is killed at home, cooling such a mound of meat will be very difficult unless the weather helps. The traditional time to kill beef on British farms was at Martinmas, in the cooler autumn weather. So for home butchering, avoid the warmer months where possible. Butchers say that to avoid what they call "bone taint" in beef (the meat nearest the bone is the first to go off) the temperature in the middle of the piece should be down to 48°F within 48 hours of slaughter.

Regulations apart, it is a risk to kill large animals at home without professional help; there is so much value to lose if things go wrong.

Preserving Meat

Freezing may be looked down upon by rugged individualists, but after all it is quick and clean and efficient until the supply of electricity fails. Freezing inhibits practically all

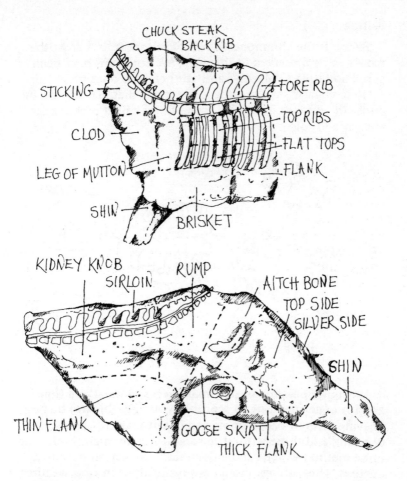

Joints of Beef

bacterial growth; they do say that lean meat freezes best, as fatty meat could be attacked by lipolytic or fat eating bacteria, as many of these can live at low temperatures. I think pork is at its best in the freezer for six months or so, and after a year the fat is not quite so good; but it will in fact keep in the freezer for years. Beef seems to improve the longer it is kept. To save space, joints can be boned and rolled. Sharp bones should be cut out, because they puncture the freezer bags.

Salting

Bacon is the commonest form of salt meat, but any other meat can be preserved in this way. Mutton hams have been cured and called "Macon" and salt beef was once plentiful; on long voyages, salt meat was the staple diet. Salting can be done dry, or wet in brine — sometimes called pickling, as vinegar and spices are also used.

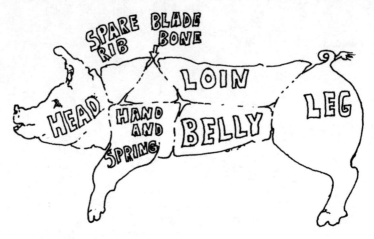

It is quite possible to salt small pieces of meat at a time and for experimenting to see what you like, this will be best. Leaving sides of bacon immersed in salt takes up a great deal of room, and those with little room to spare might prefer the brine method which can be done tidily in a clean polythene dustbin. These things should always be done in cool weather.

Dry Salting

The recommended temperature is 40°F to 42°F. Rub the piece of meat with salt for several days in succession and then leave it with a thin covering of salt, followed by a light sprinkling of brown sugar and perhaps a little saltpetre. In 10 to 14 days from the last rubbing in, the side can be washed, dried and hung up in a cool dry place. When quite dry it can be wrapped in butter muslin, but the air must be able to get to it, otherwise it will·go bad.

Some old timers recommend that to make quite sure of the cure, the thicker parts should be injected with brine, using a pickle pump. The mixture for the purpose is made from:

1 lb saltpetre
1 lb Demarara sugar
11 lb salt
4 gallons water

the ingredients are boiled for twenty minutes then cooled.

This same pickle can also be used for the wet process; the piece of meat is immersed in this solution for 10 days.

There are many regional and national variations on the basic method. One which you may like is called The Westmorland Sweet Cure.

3 lb common salt
3 lb sugar
1 lb bay salt
3 quarts strong beer

Boil these ingredients together for 15 minutes, then pour over the ham and leave for a month. Rub and turn the meat in the pickle every day. Hams should be pickled from October to March.

Making sausages is another way of preserving meat. English sausages are made in this way;

2 lbs lean meat
1 lb fat meat
1 lb breadcrumbs
½ oz pepper, pinch nutmeg and mace
¼ oz sage
salt to taste

Mix breadcrumbs with the meat and mince them all together. Add the seasoning and put through the mincer again. The sausage meat will of course not keep without freezing, but will take up less space than the bits it is made out of. Bought sausages will keep for a while only because they are full of preservatives.

If you want to keep your meat hanging up in the kitchen, the Continental type of sausage is the answer — and it is delicious too; it seems a pity the tradition never spread to Britain. The meat in these big sausages is preserved by the salt and spices and when they have been kept for a while the ingredients blend together to give a mellow flavour and texture.

235

3 lbs minced meat and fat
1 glass wine or vinegar
3 cloves garlic
spices — paprika and so on
brown sugar
1 tablespoon salt
pepper and a pinch of saltpetre

The large intestines of the pig were the traditional sausage casing, and perhaps the abbatoir would let you have some for the job; it will be far better than the bought synthetic skins which are the only alternative. Continental sausages tend to shrink as they dry, and so they are often laced into their skins. They should hang in the kitchen, the best temperature for them is about 60°F. They can also be smoked.

Smoking

This is a very old method of preserving meat; sometimes after salting. It was often used for preserving fish. An old recipe tells us that "should an added flavour be desired, the side or parts of bacon can be smoked by being suspended in a cubicle, with an opening at the top to let out the smoke, the floor being covered with peat and set alight, and the bacon kept therein from 36 to 40 hours." A month in the chimney was the standard time in old farmhouses, where only wood was burnt in the grate. Before going into the chimney the ham would be coated with bran.

Fish were lightly salted, then spread out flat and threaded on a pole, which was suspended over a smoky wood fire.

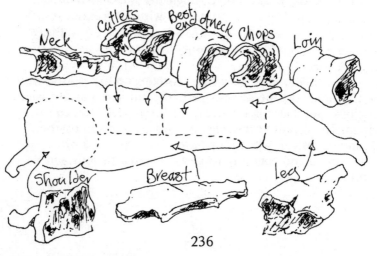

Salting Fish

Small fish under 1 lb weight can be left whole. Split the bigger ones in half and salt flat.

The fish are packed in a barrel or vat, fish and salt in alternative layers, finishing off with salt (rather like potting butter).

More salt may be needed later, to keep the strength up. When the fish are thoroughly saturated, after about three weeks in the barrel, they are taken out and washed in fresh brine. Then they are pressed flat with boards to take out excess moisture and make them thinner.

Final drying can be done in the sun if there is any, and the breeze. Flies must be kept away. Keep the fish under shade the first day to prevent burning by the sun. Six good days drying should be enough to preserve them; they can be soaked in water overnight before use.

USING SKINS

In a rural economy, nothing is wasted and the skins of the animals can always be used for something if they are properly cured to prevent their going bad. Dressing is a further refinement, to make them soft and supple and nice to wear, and is more of a skilled job.

In fact, it is not so easy to get a really luxurious finish on skin or fur at home; with the best skins, it may be wise to dry them and send them away to be dressed professionally. But everyday garments can be made quite well.

One of the favourites for home production is a sheepskin; we have several in our family, all done by different methods. It seems to me that the more time you spend on them, the better they are. Here is a simple method of curing a sheepskin. (Always start on the job as soon as the animal is skinned; this is most important for any skin at all.)

Even that unfortunate occurrence, a stillborn lamb need not be a complete loss; the skin can be used if you can get over your disappointment quickly enough to do something about it!

The first stage is to remove the skin from the carcase. To do this, cut from the head to the tail down the underside and down the insides of the legs. This will make the skin neat and

237

symmetrical. Or you can open up the belly, then pull the skin off over the head.

If it is not possible to do the job straight away, the skin must be cured temporarily by drying the pelt under cool conditions with plenty of air round it, or by rubbing the flesh side with salt and storing in a cool place. (Rabbit skins are dried by pegging out on a board).

The next stage is to soak the skins in clean water. This need only be for a few minutes with fresh skins, but if they have dried, they must be soaked until they are soft again. The best way is with clean rain water in a bucket or barrel (old butter churns make good receptacles for this).

The process continues with scouring to clean the skins and remove blood, dung and fat. The skin is put in a bucket of warm water containing 1½ to 2 oz per gallon of detergent. There it stays for an hour, and you work it occasionally, moving it about in the solution by hand. Then wring out.

Defleshing is next, and it is a tricky job. The back of the skins is scraped with a knife to remove fat or meat still left on; and the skin must not be cut. Beginners may be better with a paint scraper than a sharp knife.

After defleshing, the pelt is scoured again, using a stronger solution of detergent and another hours soaking. This should end with a clean skin.

After this comes the pickling process, to preserve the skin and fix the wool so that it will not fall out. In some recipes it is done with a solution of salt and saltpetre, but salt and potash alum, while more expensive, is recommended by the Leather Institute for a good job.

A solution is made by mixing about 1 lb of potash alum and the same amount of salt per gallon of water; the skin is soaked in this solution in an enamel or wooden container until it has fully penetrated the skin. Thin skins like rabbits and lambs will need about two or three days in this solution, while a calf or heavy sheep skin will need four to six days.

Another way is to do it dry; rub the back of the skin, flesh side, with equal parts of alum and salt, applying the paste two or three times on successive days and piling the skins together flesh side in between treatments.

Finally, the skins are allowed to drain and dry a little. Then the flesh side should be rubbed with a solution consisting of one part egg yolk and three parts of tepid water, making sure none of the egg white is there.

The skins should then be folded flesh side in and piled for two or three days, after which they should be nailed out flat to dry. They should dry gently, away from direct heat.

To soften the dressed skins they should be gently rolled between the hands and stretched over the edge of a chair or table to make them supple.

In the case of large sheep skins, patches of fat may still be seen as dark patches; they are removed by rubbing with a paste of talcum powder and white spirit, which is allowed to dry and then brushed off.

Leather

Skins without attractive fur or wool may be made into leather. Leather is made from the middle layer of the skin or hide, as it is called in larger animals. The outside layer or epidermis, with its hair follicles, sweat glands and so on must be removed at one side, and at the other the membranes and fat have to be scraped off. The middle layer is very useful to us because the pores allow for air circulation, but even so it is to some extent waterproof. Skins are first prepared, then cured and then finished. There are many different processes, resulting in the different kinds of leather.

The preparation is the same as for the home curing of skins we have just seen, but the tanning of leather or curing can be done in many different ways. There is oil tannage, using fats and oils to preserve the leather; often fish oils are used. True tanning is a vegetable process, using tannin, obtained from many plants, most often from oak bark. Then the mineral process such as we have just seen uses alum and other salts.

Rawhide This is very simple and the product has many uses on the holding, so it may be a good process to start with. Rawhide is just untanned leather — parchment is a kind of rawhide. It will dissolve to form glue in hot water, whereas tanned leather will not. Skins are washed, fleshed and so on as before; if hairy they are dehaired with a special blunt two handled blade.

The skin can be stretched on a frame, then scraped on both sides. Or the process can be speeded up by burying the hide in wet wood ash which will help to loosen the hair.

After this the skin is washed again and dried slowly; it can be worked by hand to make it supple.

Rawhide finishes up as very tough stuff and can be cut into a spiral for thongs, used to sole shoes, or for many other uses.

17 putting it all together on half an acre

For this chapter we will assume that we have a garden, plus half an acre for backyard husbandry. We have to plan how to make the best use of it.

There will first of all be room for bees in a sheltered corner, one or two hives to provide honey and fertilise the fruit trees which can be planted in the hedgerows to save space. It pays to find out what fruit grows well in the area (or ripens well would be a better way of putting it) before making a choice. Many a lover of Cox's Orange Pippin apples has hopefully planted a tree in the northern half of Britain, only to be disappointed. We have one languishing in our garden, planted by a previous owner. Cox's do better in a warmer climate. And of course the other thing to check with fruit trees is whether they can be fertilised, without which they will not set fruit. Varieties differ and even with bees on the job, only certain trees will be suitable for cross fertilisation with certain others. Both of these knotty problems can be solved by consulting Lawrence Hill's Fruit Finder.

Hedgerow trees will need protection when young, especially from goats. With only half an acre, it will probably be best to keep the field in permanent grass. Good permanent pasture can yield more than temporary leys. The grass may be poorish at first, but grazing animals will improve it by their manure and grazing, unless they are allowed to overgraze. A good stiff harrowing in winter if the grass is matted at the roots will let in the air and moisture, and better grasses are encouraged.

On half an acre, management will need to be fairly exact if we are to have a high stocking rate, ie number of mouths to feed. To get the best out of the grass it will need to be divided up. Also, the animals should be properly housed and yarded so that they can take their turn at the grass, but it is not their permanent home. In other words we need a happy compromise between a completely natural life for the animals and a stuffy indoor one. And with thought, they can have the best of both worlds. It is pleasant to see animals living entirely outside, but in a confined space they can be wretched in bad weather.

I would suggest that we could manage two goats, say a couple of weaned pigs to fatten (or a sow), about a dozen hens, and a few rabbits. A family of geese and or ducks could be additions or alternatives, particularly if water is available. Now if all this lot was to roam half an acre at will, the grass would disappear and you would be left with a bare earth compound, or a sea of mud in wet weather.

Perhaps a few hens, just enough for eggs for the family, could have free range. They would pick up insects and get the valuable protein part of their diet in this way. They must

be fenced out of the garden, but then so must everything else! Good fences are going to be essential to the whole system.

For goats on average there need to be three strands to the electric fence at 15 inches, 27 inches and 40 inches above the ground. The goats must be made aware of the fence and made to respect it.

If the hens have a house well out into the field they should be happy there and not trouble you too much. But if you would rather know exactly where they are — as can happen in a crowded neighbourhood — a fold unit will be the answer. Move the ark containing the hens onto a fresh patch of grass every day and they will leave valuable manure behind them.

Two goats were mentioned, and I suggest two for several reasons. The first is the problem of companionship; they will be happy with company and will not need you so much as a substitute herd leader. A milking goat and a young female to kid when the first milker goes dry will ensure a continuous milk supply.

The best way to keep goats in a small area of land is to have a pleasant airy house for them to live in, with a concrete yard in front. They can go out on the grass for a couple of hours every day, and the rest of the time they can skip about in the yard. This will be best from a grassland management point of view; it will also be quite good for the goats. When on the grass they can be confined to a paddock with an electric fence, or tethered. When there is more grass and you have more time the grazing period can be extended. It can be varied in country districts by a walk along quiet lanes, a very good way of augmenting a small supply of grass. Goat keepers always have an eye open for good grazing and browsing along the lanes and it is surprising how much there still is; food for free if you have the time to collect it. Walk with or without the goats; they will enjoy selecting their own, but if you are in a hurry it's quicker to dash back with an armful of your own choosing. Having been out with them, you will know what they like.

In the yards the goats can have supplementary feeding, as much variety as possible to alleviate boredom. Some branches can be tied up for them to nibble at; in the winter, roots such as turnips can be chopped in two and left for the goats and it will take them quite a time to nibble through them.

Concrete yards are not too difficult to arrange. If you do

not own the land and are unwilling to put down anything too permanent, concrete slabs or paving stones will do and they can be picked up when you move and put down somewhere else. It is a real blessing to have a yard which is clean and washable and keeps the animals off the land.

The same idea can be extended to the pigs. I suggested two weaners, but a sow could be handled in the same way. The pigsty with a yard in front, the old fashioned cottage sty, could be built next door to the goat pen, for company for the sow. Pigs and goats get on together very well and a sow on her own will appreciate the goats when she has no litter to occupy her. Some of the piglets could be sold or bartered to neighbours and the ones kept for fattening could live in a straw bale hut protected with wire, either on concrete or on a corner of land which wants digging up. Pork pigs can manage in the summer months without too elaborate a home, but those kept in winter must be off the land. With one or two acres of land one can afford to let pigs dig it up and stay on a patch more or less for months at a stretch, but on our half acre plot the grass must be kept in good condition. This means that only pigs with rings in their noses can go out to graze, and then for limited periods and in dry weather.

Pigs in a cottage sty will be fed twice a day. They will benefit from any surplus dairy products; it will probably be uneconomic to feed them full cream milk unless you are in a

244

hurry one day and have no time to deal with it. But things like whey from cheese making and skimmed milk from butter making will be very good for the pigs. One gallon of whey will replace a little under 1 lb barley meal. (Sour whey should not be fed. It is preserved by adding 0.1 per cent formalin on commercial farms).

Skimmed milk is more valuable than separated milk; the former is the result of skimming off the cream by hand for butter making and this is less efficient than a separator, so more fat is left. Separated milk has half the energy content of whole milk and can be used for cooking, and if there is enough, separated milk and a cereal mixture will fatten pigs. Thus the two goats will be providing for the pigs and even if both goats are in milk together you will still be able to make good use of the milk. When there is a surplus of milk, hard cheese can be made for storage leaving the whey for the pigs. Nothing is wasted. Everything fits in, because the pig muck will go onto a resting plot to help grow more grass for the goats. If it is composted with garden waste and put on in the winter or very early spring there will be no taint to put the goats off. The pigs will gain from exercise on the plots, a little grass will supplement their diet but they will not get much value from this on such restricted grazing. On the other hand, they and the goats will eat up any surplus garden waste or any greengrocers throw outs that you can beg for them. We fed generations of pigs on this kind of vegetable waste.

It will be best to divide up the grass into small plots and give each a rest in turn. Electric fences are the best way — more flexible and much less unsightly. Any other sort of fence on a plot as small as this will make it look like a prison compound.

In summer a three week rest will be of great benefit. Some of the parasites will be discouraged and the grass will make fresh new shoots. The pigs could follow the goats round the paddocks, i.e. graze the place the goats have just left. A milking animal always needs to have the first bite at clean sweet grass. Rabbits are kept in Morant hutches in the summer, these can graze between the goats and the pigs; but often there are odd bits of lawn and patches of grass on which to keep a rabbit hutch without using the half acre.

Alternatives Geese may be a good idea if there is enough grass for them, as breeding geese can practically live on grass, with a little grain in the breeding season and for fattening. Geese are a good thing to start with if your budget is tight. Goats and pigs are expensive, even though they eventually pay their way. Geese would be a good start and the sale of surplus goslings would perhaps earn enough money to buy the larger animals eventually. The eggs would be fertile of course, if you got a breeding set of three geese and a gander; they could be hatched under a broody hen.

Rabbits are quite a low cost alternative too. A buck and three does could graze in Morant hutches and the surplus be bartered or sold. The Commercial Rabbit Association will buy your surplus live in Britain if you can get them to their nearest collection point. This would make good use of the grass and again, earn money to put back into goats or pigs.

Vegetarians who do not like the idea of killing rabbits could try Angoras and clip the wool every six weeks. This should sell well to the growing army of hand spinners everywhere. The quantity of wool is not large, but the quality is marvellous and it can be mixed with sheeps' wool to make it go further; so if you are clever at spinning you could make up garments and have a little cottage industry on the half acre.

Hay

The half acre will be more efficient if it is not too mechanised. There will be little chance of making hay unless you know a quiet clean roadside stretch, so hay will be bought in and the need then will be for somewhere dry to store it until needed; an open barn is really best for hay because the circulation of air round it keeps it sweet. Your goats will see off two bales of hay a week and this will mean storing 100 bales or rather more for luck, at haytime. The rabbits will also need hay for the winter. Soft bedding can be made by drying summer growth in hedge banks or roadsides, and unless you do this some of the precious hay will be needed for bedding. Perhaps some straw would be useful as well, for putting in poultry nest boxes and giving a nice deep bed for the pigs. Fine hay is best in the rabbit nest.

Hay and straw at this sort of level can sometimes be obtained through barter; many people help a local farmer at hay and or harvest, when he is grateful for an extra hand, in return for their winter feed. This is good experience for a backyarder.

18 putting it all together on an acre

On an acre, the same basic principles apply as on half an acre, but with a slight difference. The animals may have a little more access to the land.

In winter, they will still need yards, particularly on land liable to poaching. But in summer the goats could be out for most of the time, with a subsequent saving of other food. The system of dividing up the grass into paddocks is very useful because if it grows quickly, a section can be cut for hay. This will be a good way of saving a surplus for the winter. And it is more interesting and better for the animals to have a frequent change of grazing.

It may be more economic to buy a sow than to keep looking out for weaners; she could come to you as a young gilt and be reared on the place or you could buy a mature sow who knows her job. A sow of one of the old breeds, used to foraging outdoors, will get about half its food requirements from good grass in early summer; comfrey could follow on and provide the same part of the sows diet until the autumn and fodder beet and potatoes, planted in a strip, could help to replace the grass part of the diet in winter. The milk products from the goats will supply another part of the ration. All this will cheapen the cost of the weaners and the sale of surplus piglets should more than cover the cost of any meal bought to fatten the ones kept for food, plus the meal part of the sow ration.

The sow will produce say fifteen or sixteen young ones in a year without too much trouble. Each porker will provide roughly twenty meals for the family, so to have pork twice a week all year you will need about five pigs reared to pork weight. Bacon pigs will give you about 150 lb of bacon, so one will supply you with 3 lb per week in a year. This sounds a lot, but the hams are something of a delicacy to bring out and share with friends, so in practice two bacon pigs might be needed if you get to like the taste of your own cure. Remember that hard outdoor work increases the appetite.

This will only account for seven pigs; there will be surplus pigs in each litter to dispose of. Unless you are onto a very good source of high protein swill you will not be wise to keep these extra pigs to fatten them. An average of say nine or ten pigs a year to sell at good strong weaner weights should bring in the barter equivalent of 3 tons of grain. Do not despise this valuable cash crop. If you need some fencing, and who doesn't, this may be a way of earning it. The profit margin will depend on management, above all on how the food can be cheapened without losing quality. Home grown food will be the answer, plus some swill.

The weaners should be sold to fellow backyarders if possible. So do some research in anticipation, and you may be able to place them all by the time they are ready. Discuss the question of castration with would be buyers; are they prepared to eat boar meat? Castration will be needed if you sell to commercial fatteners, but if a backyarder will not be keeping the pig too long, a boar will be an advantage because he will grow quickly and the meat will be lean. Tail docking is standard practice now, so you may be unacceptable because your pigs have tails!

Some people suggest that with an acre of land, you can keep a cow. This is a complicated question and in general I would think an acre is rather small. If you really want to keep a Jersey, careful plans will have to be made. Try to find some additional grazing, such as a common and by a roadside; not so easy as when William Cobbett was preaching to the cottagers. Today's suburbs are no place for a wandering cow. But of course it is easy to buy in hay and this could be part of her summer ration as well as winter feed. One could grow or buy in cabbages, potatoes, carrots — most vegetables are appreciated by cows. Someone living in an arable area might consider this. But even with paddocks, a cow will put a lot of pressure on the grazing and other enterprises might have to be cut out. A cow could be a marvellous

shared enterprise between two families, with grazing at each place in turn; this is a very good thought. The details would need to be worked out very carefully.

Text books on dairying tell you to allow half an acre per cow for grazing, on the face of it this would seem to prove that a cow can live well on an acre. But this refers to good fertile land sown down with highly reproductive grasses, the land having carried many cattle for years. Many small holdings are also fertile, but not all. But by the same token, there is nothing like carrying cattle to get the fertility into the land! So if you have dry land with no poaching a cow might be a good idea. But for a beginner, goats are probably better to start with.

Geese Where land does poach easily, geese may be the best grazers. With about an acre there could be a useful cash crop in fat geese at Christmas. Killing and dressing of poultry is a job that can be learned and it is usual to take orders early so that you have a sale for all of them before you start. The profit from this kind of enterprise depends again on the cost of the food and the work involved is such that you really need the profit! But for self sufficiency, you must need to like goose, or goose eggs. Five or six goslings a year will probably be enough for a family's needs.

When we had an acre of land, we did acquire an old

tractor, but this was rather a case of over mechanisation as far as the land was concerned. The tractor was made to earn its living by sawing up logs for sale, locally. Another justification might be a small mill run off the tractor; it would be better to buy grain and mill your own flour and meal for the animals and this would save money. An acre will give you little room for grain growing yourself, but it could be bought.

An acre is probably the place for a garden tractor plus attachments; there will be a large garden and a fair bit of cultivation to be done and this sort of machine could haul heavy material about on a trailer.

19 putting it all together on two and a half acres

There are many different possibilities with this sort of size of holding, but the best plan will depend on where you are. On a steep hillside, sheep might be the wisest main enterprise, while those on arable acres may want to have a crop rotation with temporary grass leys. Some land is best left in permanent pasture; low lying wet heavy land will yield best this way and it is always good to have a grass field near the buildings. So it all depends on personal preferences, based on what will be best for the holding. Let us look at some possibilities.

Two more things will now be possible; a cow, and hay making. The cow will increase fertility by adding her manure to the grass, thus making the grass crop increase a little each year. Surplus grass, the amount of which will vary according to the season, can be made into hay for the winter. Silage making might seem a good idea in a wet climate, but it should be remembered that to generate sufficient heat for the right sort of anaerobic fermentation, a silage heap needs to be the product of several acres and on a small scale may not be very successful; so hay would be the best bet, unless you have the time and patience to stuff your silage into polythene sacks.

Half an acre per cow for summer grazing is the ration in dairying areas under intensive management, but this may be neither possible nor desirable for backyard farming. Your land may be poorer than the average dairy farm — to start with, at least, until the heart is put back into it. Gradually

251

by keeping plenty of animals you may be able to improve the stock carrying capacity of the land. But at first it is probable that the cow will need an acre for grazing, or even more. She can have it all at once or in strips. A further acre will be needed for hay, since a cow can almost eat in a winter the hay produced on an acre, something under two tons. This meadow, as land shut up for hay is called, will be tied up by the hay crop from about March or on good land April after a quick bite off it, until it grows again after the hay, in August with a bit of luck, or September.

Then there will be two months "aftermath" grazing or a little longer in a warm dry time, a chance to rest the summer grazing patch a little. A couple of lambs would be just coming into their own as grazing machines after hay time and they could be put into the cow grazing fields.

This leaves half an acre for a pig paddock. A sow could be kept in a straw bale hut and she could rear her litter entirely outdoors. With noses innocent of copper rings, this family could turn over the paddock ready for a crop the following year; you would thus have half an acre of arable land. If you feel that the pasture needs renewing, the pigs could be taken round all the land in turn and in five years it could all be dug up and then reseeded. A crop of corn or roots could come in before it was put down to grass again.

252

Pigs that rootle tend to make the place muddy in winter, so in some climates the outdoor pig might need to come inside for the bad weather; a well made and well protected straw bale hut on concrete slabs with a yard in front would be the cheapest housing to keep her off the land. Of course if they were rung, the pigs would not dig up their paddock and could probably be left out longer, in dry climates all the year; but they should be rotated round the farm just the same. Their dung will enrich the grass, but they need to move away because otherwise the land becomes "pig sick" i.e. gets too great a worm burden. So never make the mistake of keeping the pigs on the same patch, year after year. This is one reason for having a straw bale hut; it can be burned and another one put up the next year. The alternative would be a cheap corrugated iron shelter or a wooden ark, the main thing being that the pigs need a moveable shelter.

There will probably be room to keep on the calf and rear it either as a dairy heifer or for beef; keeping calves on a small place can cause problems though. The cow needs the first bite at fresh grass, but calves should also have worm free pasture; so a separate calf paddock may be needed and this will mean buying in another calf to keep yours company. With a good milk supply, this could pay, but some bought in winter feed will probably be needed for two calves.

Sometimes cow and calf can be left together, but the calf may be difficult to wean and until it is weaned it may get most of the milk. It is more efficient if less satisfying to wean the calf from the cow at about a week and feed it on her milk, rationed out and from a bucket.

Taking away the calf from the cow has been done for so many generations that I believe the maternal instinct has been to some extent bred out of dairy cows. A beef cow will mother her calf but many dairy animals are not very interested. It was noticeable in our herd this summer that when a calf was born outside at grass, the same cow would mother it every time and the real mother often seemed disinterested. So don't worry too much about taking away the calf, they both soon forget about it. The calf accepts people as a mother substitute and will be happy with the company of a gentle dog.

Later, it may be possible to put back the cow with the calf for grazing, when they have both forgotten the relationship. This will be company for both of them.

On two and a half acres of good land there may be room

for a goat and a cow or perhaps two small cows. This would keep you in milk for all of the year, with a surplus sometimes which could be made into cheese and butter and used for rearing pigs. It may seem a little greedy to want so much milk, but with just one cow there will be an average of two months in the year when she is dry and you have no fresh milk. With a regular supply your animals can depend on it as well.

After the best grazing has been taken by the milking animals there should be a little grass left behind. Sheep as mentioned before would fit in well here. It may be a good idea to get a couple of orphan lambs in Spring and rear them on the bottle, on cow or goats milk. On the half acre plot dug up by the pigs there would be many possibilities for crops to fit into the overall plan. This plot will help to alleviate the headache of winter keep. On most places, the grass will not grow very much in the winter and something else will be needed to fill in with the hay ration. A strip of several kinds of crop might be better than half an acre of one thing.

One consideration must be the expected yield. An average yield for kale might give you six or eight tons on the half acre plot, which sounds a lot to get through! A frost hardy kale such as thousand head will give cows, goats, pigs, hens and rabbits green food right through until the grass grows in the Spring. Once the crop is established kale seems to be very hardy and to suffer from few diseases; keeping the weeds down might be a problem in early summer. The kale can be used from about September, either grazed or cut and carted to the animal, which is preferable especially in wet weather.

A Jersey cow will eat about 50 lbs of kale a day. This means she can get through about two tons in the course of the winter. Feed her a little hay before the kale in the mornings, to prevent digestive upsets.

Sheep do well on kale. A ewe or a strong lamb (ewe orphan lambs could be kept for breeding the next year) will eat about a hundredweight a week, so two ewes will dispose of over two tons in the winter.

Chopped kale can be fed to pigs and 7—8 lbs kale will replace 1 lb meal. The rest of the crop can be cut for the poultry and rabbits, and any left over stalks can be composted.

Kale is highly digestible and rich in protein and also in calcium. But it is low in trace elements, and so a mineral lick should be provided. It is particularly deficient in iodine and it should be remembered that a goat has a bigger thyroid gland than a cow and therefore a greater need of iodine. Seaweed meal will provide the extra.

Half an acre of potatoes is another useful possibility. Raw potatoes have a laxative effect, and may cause scouring in large amounts, but they are easy to weigh and ration.

The yield on half an acre might be about three tons — or more on good land. Some of these — they would be main crop varieties — would be accounted for by the household, but all the animals would benefit as well.

Cattle can eat potatoes raw, and the Jersey cow might have about 20 lb per day or about a ton during the winter. They must be stored properly away from the frost, of course.

Sheep can have 4 lb per day each.

For pig feeding, potatoes really come into their own. Boil them up for pigs and poultry to get the best value out of the starch. 4 lbs of potatoes will then replace 1 lb of meal. A bacon pig can eat up to 14 lbs of spuds, and any separated milk or whey will almost complete the diet. Geese can be fattened on cooked potatoes, dried off with a little meal.

Fodder beet, turnips, carrots, cabbages — all these would be suitable crops and a strip of each would give the animals a varied diet through the winter. Variety would also ensure you against a failure of one crop. Work out the capacity of your stock to get through whatever crops you think of growing; it would be wasteful to grow more than you can deal with unless a surplus could be bartered for something else, or sold, as perhaps carrots or potatoes could.

The cereal enthusiast may be hankering after half an acre of wheat. This would be possible, but bird damage is considerable in small patches of cereals and the actual cash value of a cereal crop is one of the lowest of all the farm crops. You tie up an acre of land for a year with autumn sown corn and for the whole growing season in any case, for a yield of say a ton and a half of corn at best. The value of grass would be far greater on the same land. Then there is the problem of harvesting a small amount. In an emergency we would no doubt do it, but to my mind it is more sensible to buy a bag of grain at a time from a larger farmer.

255

20 putting it all together on ten acres

With ten acres we have a small farm. Once again, the possibilities are very varied and depend on climate and soil more than anything else. Capital available will also come into it, of course, and it may sometimes be wise to go very steadily at first until some capital can be accumulated. But on ten acres, the grass will continue to grow and it must not be neglected. As a last resort, one could graze a neighbour's cattle or sheep for a season.

This is not so defeatist as it sounds; it has been the salvation of many a beginner. There are a few simple rules which it will be well to observe.

First of all, make sure the fences are stock proof; sheep proof fences are a lot tighter than ones which will hold cattle comfortably so don't take sheep unless you are sure they will stay at home. Mountain sheep almost never do! A little fencing before you start — and this applies to your own animals as well — will be worth a lot afterwards because once they get the idea, animals will force their way through otherwise stock proof boundaries. Don't put it into their heads.

Then, make sure of your agreement — how much per head, and the exact length of stay. Be quite sure when they are going off again. And don't be tempted to undertake too much if you are not sure what the grass will stand. Don't let anyone pasture their stock on your land continuously for anything approaching a year; a few months at a time are best. A grazer can if he has the land all the time establish an agricultural tenancy and you may not be able to get him off — ever.

The advantages of letting the grazing, if you are new to the farm, are several. Firstly, the land is bringing in certain profit and whether the animals pay their owners or not, you will still get your rent. The time will allow you to get your ideas sorted out, to get to know the farm and the soil and to work out a plan of action, always easier on the spot than anywhere else. Try to get an arrangement where you are actively concerned in looking after the animals, going round them and so on. This will be valuable experience when you come to have your own. Question the owner, find out what he is doing and why, try to ascertain what he sells them for! You will gradually build up local knowledge and contacts in the area and you will see at first hand how long it takes a certain number of animals to eat down your grass. Then, a field at a time, you can begin to take over your land. And all this time the land will be earning money; the prices obtained for summer and winter grazing are generally high.

This is of course a temporary expedient; you will be wanting as soon as possible to be farming it on your own. On ten acres of average land, there is the chance to work out a proper rotation of crops and to grow temporary grass on each field in turn, ploughing it in after a few years and re-seeding it later with good grass.

257

The first crop might — according to the area — be wheat after grass, followed by roots and mangolds, fodder beet or swedes. After this might follow a crop of barley or dredge corn before you sow the field down to grass again. If you ploughed up two or three acres every year there would still be an appreciable amount of grazing left.

On thin soils, wet places or upland holdings it will be unwise to plough up land and it is worth remembering that an all grass farm is good for the land and easier to work. If I had a ten acre farm and a job as well, I would want to leave the land down to grass except for a garden. Haymaking will give you enough field work! a diversity of animals will ensure that the land is not exhausted by one crop. On the other hand, if the land is good but the pastures poor, you will want to take the plough round the farm, as they say, to get better grasses growing.

A couple of cows plus followers should keep you in milk for the whole year, with a surplus for animal rearing. The choice of breed is wide; Jerseys if you want a lot of cream, but there will be room for bigger cows. For an all dairy enterprise the Ayrshire is a very hardy, sensible little cow and the milk is good for cheese making. Some people prefer to keep an animal with more beef potential and go for the Welsh Black, particularly in upland Wales. The Welsh Black used to be a good dual purpose cow and fifteen years ago there were many herds of this breed in North Wales which were used for milk selling as well as calf rearing — a thousand gallons per lactation was quite common. But it seems to me that they have been keener on the beef side lately. But any types of beef breed, including the European ones are extremely docile and would be able to produce enough milk for a calf plus the household, at least.

A Charollais cow for example should be able to pull implements or a cart and give milk — a pleasant thought.

A compromise which we use is to have dairy cows, Ayrshire or Jersey, and use a beef bull such as the Hereford so that the calf is fairly beefy.

Most people think that ten acres should be used for more than food self sufficiency; it should be able to contribute to the family income — to pay its way. It will hardly make a living in organic farming without some supporting cottage industry such as basket making or pottery, but it should leave a surplus for cash expenses as well as providing most of the necessary food.

Whatever the enterprise chosen, one must be business like. Keep strict accounts — they need only be simple. Basically what you will need will be an account of all money spent on the farm, and receipts to prove it, on the one hand. On the other there will be a list of all the money earned by the farm. It will pay to get an accountant to present these for tax purposes because without doing things properly you could end up paying more tax than you need. Hitherto in talking about backyarding we have assumed that little money will need to change hands; but a ten acre place will have its own expenses and these can only be set against tax if the thing is done properly.

What sort of an enterprise you choose will depend on personal preferences and skills, but also on what sort of a place it is and what sort of markets are within reach. Hill areas are often remote, but in summer they can cater for tourists. One way of selling your surplus produce in a tourist area is to offer bed and breakfast to travellers. As awareness of the organic movement grows, the offer of properly grown food will no doubt be a special attraction. If you are a sociable family this can be a pleasant way of earning money. Some families new to country life can feel a little isolated at times, and catering gives a contact with people which can be welcome. So if you have just a little more milk, eggs and other food than you need, this is a way of turning it into a cash crop.

Near a town, there are other possibilities. Market gardening has become one long chemical spray; manpower has been replaced by chemicals. The crops are grown so professionally and quickly that none but an expert can compete on their terms. But there are now people who want to buy organic produce. There are organisations which will help you to find a market for it. One can always start in a small way and see how it goes.

Another way of using your ten acre piece is to get geared up for selling milk. Goats milk will usually be best, because there are no restrictions on its sale. And also, the capital outlay on cows is enormous compared with that for goats, and more than 10 acres would be needed for a reasonably sized enterprise.

Retail deliveries of liquid milk will pay best if you can find the customers and they all live close together, but it can take too much time. An alternative is to find a hospital interested in buying goats milk; it is recognised as being very helpful in the treatment of many diseases. If you are rather far from civilisation, the making of yogurt might be a good way to sell the goats milk. Health food shops will be pleased to find a source of goats milk yogurt and in this form it will keep for a few days longer than milk, so you need not make daily deliveries. This is the easiest and cheapest of all the dairy products to make.

In a really hilly area the cash crop might have to be sheep, with the ewes away-wintered on another farm lower down the hill. This is an old tradition; the hill farms sell store lambs to the lowlands for finishing to fat lambs and this plus the fleeces could be your crop. More ewes could be kept if you have access to moorland or common land. Or, to consider the other side of this, you could winter sheep for other people after grazing the area in the summer with your own stock. The wintered sheep would tidy up the fields and make their contribution to fertility. With a bunch of ewes it might be easier to keep a ram.

Hens are tricky subjects for a cash crop; the battery systems produce eggs in such quantities that the price is kept down and you will need careful accounting to make sure they are not running into the red. With home grown food the picture should be rather better, and it should be possible to get a better price for authentic free range eggs. Geese are usually a good proposition where there is plenty of grass but the drawback here is their rather mean nature; small children should not be allowed near geese. I have dreadful memories of our geese biting the legs that fed them.

Cash crops, whatever they are, are only part of the overall plan. A little cash may make the food production easier; it may provide power, a horse or a tractor, to help with all the jobs.

On our ten acre holding it should be possible to arrange some fish culture ponds, especially if there is a source of water available. Since ducks and fish go well together they could be included in a mixed programme. Raising a dyke to keep in the water and setting the whole thing out will obviously take a lot of time and effort. But it should not be too expensive and it should prove to be really worthwhile.

Bearing in mind that protein production per acre with fish can be twenty times that of animals, here is a basically efficient enterprise that fits in well with all the others.

The problem of producing cheap eggs and pig meat is really one of finding the necessary protein. Now if the small fish, surplus ones and offal can be fed to pigs and poultry as their protein, with home grown grain supplying the rest, that problem is solved. But this is not all; there is a two way system. Fertilisers to increase the pond production need not be bought — the pig and poultry manure can be used.

21 regulations

British law is notoriously vague and while some of the regulations which concern us as backyarders are quite clear, others are not. This chapter is intended as guidance, but if in doubt about your legal position consult someone who really knows. Solicitors can be very helpful; other sources of help are the Citizens Advice Bureaux and the police. The Ministry of Agriculture will advise you of their regulations and the most relevant department will be Animal Health.

Ignorance of the law is of course no defence. Intent on doing your own thing, it is easy to forget about regulations; but it is much simpler in the long run to know the rules and to keep them in this overcrowded world. It saves much time looking over your shoulder to see if "they" have caught up with you yet!

Another good reason for knowing where you stand is the officious individual who will come along and watch your efforts and then tell you that you are breaking the law. Some people love to do this, especially to children, and the best defence is to know the law yourself.

Rules About Animals

Cruelty. The Protection of Animals Act applies to all animals not at liberty. It is an offence to ill-treat an animal or to cause it unnecessary suffering. No reasonable being would wilfully hurt an animal, but it is easy to be careless or forgetful. Keeping animals is a heavy responsibility and I mention the law as a reminder of this.

262

Keeping Animals. Keepers of animals are responsible for any damage or injury caused by the animal, with some exceptions; if your children are under sixteen, you are liable for their animals.

Animals are divided legally into dangerous and harmless species, and all domestic animals are supposed to fall into the "harmless" class. This is a good thing, because you are liable for trouble caused by a "harmless" animal only if you know or ought to know that it is likely to be dangerous. You must keep your animals under proper control so that they don't stray — especially onto the highway.

Cats are mavericks — the owner of a cat is not responsible for any damage it may do because it is accepted that nobody can really control a cat. Bees, it seems, can get you into trouble if they sting your neighbours, if it can be proved that you have not handled them properly.

Animal Health Regulations

You must by law keep an Animal Movement Register of the movement of all cloven-hoofed animals on and off the premises. This includes pigs, cattle, sheep and goats. Rule out a notebook, *Date. Animals. Premises from which moved. Premises to which moved.*

In the case of an outbreak of disease the police or the MAFF or the local authorities will want to see your record — always unexpectedly!

Some diseases are listed as "notifiable". Among these are swine fever, SVD, foot and mouth, anthrax, Teschen Disease, fowl pest and a new one, EBL (Enzootic Bovine Leucosis, a transmissable virus disease of cattle so far confined to Europe). The owner of an animal suffering from a notifiable disease must inform the police, but in practice you will have called in the vet to make sure and he will sort it out for you because it is also his duty to report these diseases.

The Ministry of Agriculture make Orders to prevent the spread of infectious disease and if an outbreak occurs they impose movement restrictions. At the time of writing there are restrictions on the movement of pigs in Britain in an effort to get rid of Swine Vesicular Disease. Pigs may not be moved from one place to another without a permit, unless they are going for immediate slaughter. This permit is usually

obtained by the owner of the pigs. It is obtained from the
police or from the county council special officer. Once pigs
have been moved onto your premises, you will not be issued
with a permit to move any off again for three weeks.

Sheep dipping is at the moment compulsory again for the
control of sheep scab. The special officer of the council also
deals with this and he will tell you what regulations are in
force. In general, sheep have to be dipped between certain
dates. If you use one of the organic sheep dips they will have
to be done twice at eight day intervals — see the Backyard
Sheep Farming Book for details of these.

Swill Boiling

The regulations about this are now rather complicated.
Briefly, waste food fed to pigs must be boiled, and kept
boiling for at least one hour. The premises have to be arranged
so that raw, incoming swill is kept away from the cooked
swill to prevent any recontamination. The only way a backyarder can feed waste food to pigs without boiling it is to put
them on a vegetarian diet. No meat, bones, offal, eggs or
hatchery waste must be fed and this includes cooked meat
such as sausage meat. This is very important and the rule is
not one to be dodged. It would be very inconvenient for
many backyarders if this outlet was removed and if people
abuse the privilege it can only make the rules tighter. You
could of course have enough pigs and waste food to justify
boiling; in this case consult the MAFF Animal Health Department for advice.

Food From The Wild

Under the Theft Act of 1968 it is not stealing to pick
mushrooms or fruit or foliage from a plant growing wild on
any land unless it is done for sale or other commercial
purposes. This law might have been framed with backyarders
in mind. It excludes wholesale taking of nuts, berries and so
on just to sell them, but it allows us to gather things for our
own use, although not, for example, to make jam for sale.
The plants must be wild; if somebody has planted them it is
stealing — e.g. to take a turnip without permission. The law
does not allow us to dig up plants; they must be left so that

they will survive. If you want to propagate wild plants in the garden, collect the seeds.

Likewise, wild animals are not supposed to be anyone's property and they are not stolen if taken unless they have been "reduced into possession"; that is they are although a wild species, kept in captivity. But of course if you take a wild animal without the permission of the owner of the land you are poaching — or even if you just trespass with this in view. Poaching is a more serious offence if there is a gang of more than five and if it is night.

A gamekeeper can arrest poachers at night, but by day only if the poacher will not give his name and address and refuses to leave the property. The police can stop suspected poachers at any time.

Trespass is not a crime except in a few cases, such as on railway property or Ministry of Defence land. But the owner of land has the right to enjoy it in peace. The law may affect backyarders from either side, when out for wild food or when at home repelling invaders. If you get tired of repeated trespass by someone over your land, you can sue for damages, but it is not very usual and there would not be much compensation unless real damage had been done. It is more usual to get an injunction, an order from the court forbidding further trespass. Any one defying this would be in contempt of court. You can keep guard dogs, but should put up warning notices. It is a sad fact of our times that people are finding it necessary to guard their gardens in crowded areas. There are now special restrictions on guard dogs on premises other than agricultural land, houses and gardens, but our activities would come into these categories.

If you trespass on someone's land you are there at your own risk — except for children, for whom the owner of the land has a responsibility.

Shooting

For a rifle you will need a firearm licence from the police. They will want to know why you need it and where you intend to use it, so the first thing is to get permission to shoot if you have no land of your own — and the police will check any address that you give. Rifles are dangerous and it is not easy to get a licence.

Shotguns are rather easier in country districts, being part of the tradition, but the police will keep an eye on you even so and you will need a certificate. Renew gun licences very promptly, otherwise prosecution may follow. Game licences are needed for shooting game birds and deer, but even so there are limited seasons of the year in which game may be shot. They are as follows:

Grouse   August 12 — December 10
Partridge   September 1 — February 1
Pheasant   October 1 — February 1

Even in the season, Sundays and Christmas Day are close days.

DEER Males are not killed May 1 — July 1.

Females March 1 — October 1.

This is except for killing to prevent suffering or serious damage to crops. With all the licences you also need the permission of the owner of the land.

The following birds may only be killed outside the close season, 1st February — 31st August. (Except when they are below high water mark; when the close season is 21st February — 31st August.)

Grey Lag Goose, Bean Goose, Pink-Footed Goose, White Fronted Goose, Mallard and wild duck, teal, widgeon, Gadwall, Shoveler, Tufted Duck, Pochard, Pintail, Goldeneye. Also Moorhen, Golden Plover, Curlew, Whimbrel, (the Stone Curlew is totally protected).

Unprotected birds.

House sparrow, starling, rook, woodpigeon, great black-backed gull, herring gull.

Mammals to be Spared

The hedgehog used to be a delicacy. While not protected it is disappearing and should be conserved.

The badger — now protected.

The red squirrel is rare.

The Dormouse is rare.

Fishing

British people have the right to fish in the sea up to 6 miles from the coast, and usually also in river estuaries and tidal waters, whereas non-tidal stretches of river are usually private and the fishing rights kept by the owners. To find out which waters are tidal, look at the Ordnance Survey Map. Tidal waters are edged with a black line, non tidal waters with a blue line. If you fish in private waters without permission you may be prosecuted.

There are close seasons for certain fish, when it is illegal to take them. All these close seasons are arranged with the breeding season in mind, so that the numbers may be kept up to a reasonable level — in other words, for conservation, and to be respected.

Fishing Close Seasons

Salmon	September 1 — January 31
Trout	September 1 — February 28
Freshwater fish	March 15 — June 15

Different close seasons may be established by bye-laws of the river authority if they feel that there is a need; obtain details from the local water authority or from the owner of the water.

Salmon and trout may not be caught by commercial means from 6 am on Saturday to 6 am on Monday. The river authorities issue licences for salmon and trout fishing — a rod licence.

Sea Water Fishing

These are the National Federation of Sea Anglers' minimum sizes; greater than the legal sizes, but it would be best to throw back any smaller than these.

Grey Mullet	13 inches	Flounder	8	inches
Codling	12 inches	Gurnard	9	inches
Pollock	10 inches	Lasser Sand Eel	9	inches
Whiting	10 inches	Wrasse	9	inches
Bass	15 inches	Sea Bream	9	inches
Plaice	10 inches	Red Mullet	13	inches

The public has a general right to navigate over tidal waters, but rivers are private, except in certain places like the Thames.

Commons

It may be useful to be able to prove a right of common

and many people try to do so, but it can be difficult. It can be done by producing a written document; for example, the deeds of our little farm in Wales stated that we had the right to pasture sheep on the common land above the farm. But while we were there this right had to be registered under the Act of 1965, and if this had not been done, the right would have been lost. So a holding with registered grazing rights is now the only way to get them.

There is one other way; to prove that you live on the land of the same manor in which the common is situated and that you or your predecessors have exercised that right "from time immemorial". Not everybody has mediaeval records, so in practice this is taken to mean for a number of years; so it is only useful if rights are being taken away from you. It does emphasise that if you have common rights, you should hang onto them and use them.

Some usual common rights; Common of pasture — horses, cattle or sometimes geese are grazed on village greens. In some areas such as the New Forest there is the right of Pannage — pigs can be turned out on wooded common to eat the nuts.

Common of estovers — this is the right to take wood for the repair of the house and for fuel. It was probably responsible for some of our deforestation.

Common of Turbary — the right to take turf or peat for fuel.

Common of Piscary — the right to fish in streams or ponds on the common.

Common in the Soil — the right to take sand, gravel or stone for household use.

Once again, all these rights are basically for self-sufficiency and not for commerce.

The Coast

The foreshore is the strip of land between the high and low water marks. Most belongs to the Crown, but some has become private property. So that apart from boating, most uses of the shore are by implied permission of the Crown, which can in theory be withdrawn. So we have no absolute right to collect driftwood, seaweed or shells; but by long tradition, we do these things.

Selling Produce

Bartering of surpluses is usually preferable to selling because it dodges several issues. In the first place if you sell for money you may make a profit, which will be taxable. Gains arising from barter will no doubt be taxable where they can be discerned, as my legal adviser tells me; but fair barter is an exchange of goods of equal value and gain does not come into it. But profit arising from sales will need to be accounted for.

Then there is the legal aspect of selling, one extremely complex. The Sale of Goods Acts and the Trades Descriptions Acts have to be observed, and with food the Food and Drugs Acts as well.

Milk is easy if it is goats milk — no restrictions; but to sell cows milk your premises have to be registered with MAFF and come up to standard, as has the water supply. The Milk and Dairies Regulations are a good basic guide to clean milk production and it is well to follow them whether you sell or not. But to sell by retail you need an untreated licence and in theory the milk is bought from you by the Milk Marketing Board and then sold back to you again with a levy on it. Cheese and butter selling will also involve registration.

Eggs are easy to sell — you can sell them to householders but not to shops.

Killing animals for home use is legal. If you are going to sell your surplus animals like pigs and sheep, they have to go to the abattoir to be killed on proper premises by a licenced slaughterman. Selling meat from any animal killed on your premises means you must have your premises inspected and licenced, which is out on a small scale.

If you sell anything by weight (eg tomatoes) the scales should be inspected by the local authority from time to time.

Land Ownership

If you are in possession of land for twelve years or more undisturbed without paying rent, you may be able to establish ownership. There have been cases where payment of rent has lapsed and the real owners have been unable to get their property back. But this is a very involved law and it is unlikely to be useful in practice.

Boundaries

The deeds of property you buy will describe the boundaries in words, or the plan may define the limits; the words are usually the legal definition. The deeds may specify ownership of walls and hedges but not invariably. It pays to walk round and sort them out for yourself. Locals of course will know to whom the boundaries belong. You will find yourself responsible for the upkeep of some of them, but not all.

There are a few general rules for cases where ownership is not clear. Walls are presumed to belong equally to the owners on each side. A fence belongs to the person with the stakes on his side. Where a stream forms the boundary, the hedge or fence on your bank is your property, because the middle of the stream is the boundary.

Where a hedge or fence is bounded by a ditch the boundary is presumed by the hedge and ditch rule to be at the far side of the ditch. It is presumed that a man dug the ditch on his own land, threw up the bank thereby, and planted the hedge on top of the bank.

Even where a poor fence or hedge belongs to your neighbour you should fence to keep in your own stock. It may be policy to keep out your neighbours animals! In sheep country this can be difficult and expensive and the cause of many a rural quarrel.

Water You do not own the water in the stream which happens to run through your land. It can be used in a reasonable way, for ordinary household purposes. But if it is to be extracted in large amounts you will need a licence from the water authority, for example for irrigation.

reference

General Books

The Complete Book of Self Sufficiency by John Seymour (Faber 1976)
Fat of the Land by John Seymour (Faber) his earliest and best.
The Countryside Explained by John Seymour (Faber 1977)
Brave New Victuals by Elspeth Huxley (Chatto and Windus) lowdown on bought food.
The Living Earth and the Haughley Experiment by E. B. Balfour (Faber reprint 1975) the principles of organic husbandry.
Cottage Economy by William Cobbett. Facsimile reprint by Landsman's Bookshop 1975.
The Farming Ladder by E. Henderson (Faber). Good sound principles.
Farmers of Forty Centuries by King (out of print) Organic sense (Cape 1926)
The Owner Built Homestead by Ken Kern (Prism Press)
The Owner Built Home by Ken Kern (Prism Press)
Farming and Wildlife ed Barber, (RSPB)
Manual of Nutrition MAFF, HMSO 1976.

Soil Association booklets — they are all useful.
Smallholder Harvest.
Self Sufficient Smallholding.
Use Your Weeds.

Agricultural Notebook McConnell. Facts and figures (Iliffe Books).
Observers Book of Farm Animals by L. Alderson (Warne)

Buildings

Build Your Own Farm Buildings by Frank Henderson
The Owner Built Home by Ken Kern
Shelter Belts and Windbreaks by J. Caborn

271

Poultry

Backyard Poultry Book by Andrew Singer (Prism Press)
Natural Poultry Keeping by Jim Worthington (Crosby Lockwood)
Ducks and Geese HMSO Bulletin 70

Rabbits

Backyard Rabbit Keeping by Ann Williams (Prism Press)
The Private Life of the Rabbit by R. M. Lockley (Deutsch)
A Manual of Rabbit Farming by M. Netherway (Watmoughs 1974)
The Domestic Rabbit by J. C. Sandford

Bees

Backyard Beekeeping by Wm. Scott (Prism Press)
MAFF Bulletins: Beekeeping no. 9
                Swarming of Bees no. 206
                Diseases of Bees no. 100

Advisory leaflets (free)
                367 National Hive
                445 Smith Hive
                468 Modified Commercial Hive
                549 Langstroth and MD Hives
                412 Feeding Bees

NB: See other leaflets for other subjects — MAFF publications are useful.
Making Mead by B. Acton and P. Duncan

Goats

Goat Husbandry by Mackenzie (Faber) the classic
Keeping Goats by Elizabeth Downing (Pelham)
Observations on the Goat by M. H. French (FAO, Rome)
The Backyard Dairy Book by Len Street & Andrew Singer (Prism Press)

Pigs

Backyard Pig Farming by Ann Williams (Prism Press)
Pig Husbandry by J. Luscombe (Farming Press)

Sheep

Backyard Sheep Farming by Ann Williams (Prism Press)
Profitable Sheep Farming by M. McG Cooper and R. J. Thomas (Farming Press)

Cattle

The Family Cow Dirk van Loon (Garden Way, Charlotte Vermont USA)
The Herdsmans Book by Ken Russell (5th Edition, Farming Press)
Backyard Dairy Book (Prism Press) see goats.

Fish

Backyard Fish Farming by Paul Bryant (Prism Press)
Fish Farming by C. F. Hickling

Health

Herbal Handbook for Farm and Stable by Juliette de B. Levy (Faber)
Herbal Handbook for Everyone (Faber)
The TV Vet Books (Farming Press) Guides with pictures to treating cattle, pigs, sheep, horses.

Technology etc

Cloudburst 1 and 2 handbook of Rural Skills — all kinds of gadgets and ideas: Cloudburst Press Ltd., Box 79, Brackendale, British Columbia VON 250 Canada.
In Britain, may be available from Compendium Bookshops, London.
Tools for Agriculture by John Boyd 1976. Buyers guide to low cost agricultural implements — worldwide.
Also a magazine, *Appropriate Technology* published quarterly. Both from Intermediate Technology Publications Ltd. 9 King Street, London WC2 8HN.

Survival Scrapbooks. Energy, tools, food. Series of manuals "for survival in the 20th century dissolution". Full of ideas and addresses.

Radical Technology 1976 by Undercurrents, 29 King Street, London WC2E 8JD. Food, shelter, tools, materials, energy etc.

Plants

Comfrey Past, Present and Future by L. D. Hills (Faber)
Crop Husbandry by Lockhard and Wiseman (Pergamon 1975)
A guide to Wild Plants by Michael Jordan (Millington Books)
Food for Free by Richard Mabey (Collins)
Enjoy Your Weeds by Audrey Wynne Hatfield
Eat the Weeds by Ben Charles Harris (Barre Publishers, Mass. USA 1972)
The Weed Cookbook by Adrienne Crowhurst (Lancer Books, New York)

Crafts

Your Handspinning by Elsie Davenport
The Use of Vegetable Dyes by V. Thurston
Willow Basket Work by A. G. Knock
Country Bazaar by Pittaway and Scofield (Fontana/Collins)
The Shell Book of Country Crafts

WHERE TO GET SUPPLIES

Self Sufficiency and Smallholding Supplies (Catalogue available) all the things that are hard to get; hand tools, dairy equipment, small grinders, rennet etc.
The Old Palace, Priory Road, Wells, Somerset, England.

Small Scale Supplies "everything for the self supporter". Books, greenhouses, cheese moulds poultry equipment etc. Widdington, Saffron Walden, Essex.

Dairy Equipment

J. J. Blow Ltd., Old Field Works, Chesterfield, Derbyshire, England.

Cream separators: *Alfa Laval Ltd.*, Cwmbran, Mon. Wales.
Foothills Creamery 4207 16th St., Calgary, Alberta, Canada.

Goat Equipment

Fred Ritson, Goat Appliance Works, Carlisle.

Dalton Supplies Ltd., Nettlebed, Oxfordshire, England.
Eartags, weighbands for estimating pig weights, kicking bars
for erring cows, etc.

Rennet

Fullwood and Bland Ltd., Ellesmere, Shropshire, England
Chr. Hansen Ltd., Basingstoke Rd., Reading — also dairy
cultures, e.g. yogurt culture, cheese starters, butter flavours,
annatto.
In USA: *Chr. Hansen Lab. Ltd.*, 9015 West Maple St.,
Milwaukee, Wisconsin 53214

Rabbit and Poultry Equipment

Eltex Ltd., Eltex Works, Worcester, England

Beekeeping Equipment:

E. H. Thorne Ltd., Beehive Works, Wragby, Lincs.
Walter T. Kelley Co Inc. Clarkson, Kentucky USA

Other Requirements

Rossendale, 25 Lumb, Rossendale, Lancs.
Fencing — electric mains.

Thompson and Morgan, London Road, Ipswich, Suffolk
Seeds including TREE seeds, unusual sorts, backyarding
equipment and fish farming gear — good catalogue with information on growing.

Cade Horticultural Products Ltd., Streetfield Farm, Cade
Street, Heathfield, Sussex.
Small amounts of farm seeds.

Fish

The Cotswold Carp Farm, Broadlands, Bourton on the Water, Gloucestershire.

Humberside Fisheries (Ken Ryder) Cleaves Farm, Skern, Humberside.

Thompson and Morgan, Ipswich.

Waterwheels and machinery put right — qualified millwrights *Colman and Unwin,* 91 Victoria Road, Diss, Norfolk.

Craft Equipment — spinning and weaving

Eliza Leadbeater, Rookery Cottage, Dalefords Lane, Whitegates, Northwich, Cheshire.

Malcolm MacDougall, Grewelthorpe, Ripon, North Yorkshire

Frank Herring & Sons, 27 High West St., Dorchester, Dorset

Handweavers Country Style, Norfield, Vermont 05663 USA

The Makings, 2001 University Avenue, Berkeley, N. California 94704, USA

Buildings

Sectional wooden buildings: *Browns of Wem,* Four Lane Ends, Wem, Shropshire.

Farmstead Developments Ltd., The Station, Ashwell Thorpe, Norwich, Norfolk.

Small Livestock. Goats, Poultry, Ferrets, Pigeons, Rabbits etc.

Abbott Bros., Thuxton, Norfolk.

EDUCATION

Westdean College, Chichester, Sussex — spinning, butter and cheesemaking, all kinds of skills taught.

Spinning — *Eliza Leadbeater* teaches it — see equipment.

Courses on self sufficiency — *Aylesbury College of Further Education,* Stoke Mandeville, Bucks.

Short course on organic husbandry: *Soil Association,* Walnut Tree Manor, Haughley, Stowmarket, Suffolk.

Working Weekends on Organic Farms — a good idea for beginners to find out what they would like to do Contact *Soil Association*.

For short privately run courses, see advertising section of the magazine *Practical Self Sufficiency*.

USEFUL ADDRESSES

Some useful organisations

The Soil Association, Walnut Tree Manor, Haughley, Stowmarket, Suffolk.
Advice on organic gardening and farming and a Journal.

The Henry Doubleday Research Association, Convent Lane, Bocking, Braintree, Essex. Advice on organic gardening and farming and a Newsletter.

Both these are now world wide organisations and give a great deal of help to people like us. Well worth joining.

In USA: *Organic Gardening and Farming*, Emmaus, Pa 18099

British Beekeepers Association 55 Chipstead Lane, Riverhead, Sevenoaks, Kent (Helps amateurs and professionals)

Commercial Rabbit Association, Tyning House, Shurdington, Cheltenham, Glos. (Helps with selling)

National Sheep Association, Jenkins Lane, St. Leonards, Tring, Herts. (Jacob Sheep Society at same address)

The Jersey Cattle Society of the UK
Jersey House, 154 Castle Hill, Reading, Berkshire.

American Jersey Cattle Club
2015J South Hamilton Road, Columbus, Ohio 43227

National Pig Breeders Association
49 Clarendon Road, Watford, Herts.

British Goat Society
Rougham, Bury St. Edmunds, Suffolk.

Rare Breeds Survival Trust
The Ark, Winkleigh, Devon.

Countrywide Livestock
Market Place, Haltwhistle, Northumberland (for rare breeds).

The Organic Register (Voluntary organisation to promote
sale of produce)
Old Rectory, Withington, Glos.

The Alternative Technology Centre,
Machynlleth, Wales.

Her Majesty's Stationery Office
(Source of some excellent bulletins and leaflets)
York House, Kingsway, London WC2

Periodicals:

Practical Self Sufficiency pub. alternate months by Broad Leys
Publishing Co., Widdington, Saffron Walden, Essex.
A very useful publication.

Beecraft (monthly) from 17 East Way, Capthorne, Sussex.

Fur and Feather (rabbits etc.) Watmoughs Idle, Bradford.

Organic Gardening and Farming (magazine) Emmaus, Pa. 18099
USA

INDEX